Strategic Engineering

I would like to thank Caroline Arcade, a sociologist, collaborative project engineering consultant, and an advisor in integral ethics, for the valuable contributions she made to the writing of this book

Innovation and Technology Set

coordinated by
Chantal Ammi

Volume 11

Strategic Engineering

Jacques Arcade

iSTE

WILEY

First published 2021 in Great Britain and the United States by ISTE Ltd and John Wiley & Sons, Inc.

ISTE Ltd
27-37 St George's Road
London SW19 4EU
UK

www.iste.co.uk

John Wiley & Sons, Inc.
111 River Street
Hoboken, NJ 07030
USA

www.wiley.com

Library of Congress Control Number: 2020951813

British Library Cataloguing-in-Publication Data
A CIP record for this book is available from the British Library
ISBN 978-1-78630-642-5

Contents

Foreword

In this book on strategic engineering, Jacques Arcade seeks to extend and deepen his expertise in foresight and strategic management in an innovative methodological approach, bringing together vision, anticipation, and action in increasingly complex and changing, even fluid, environments. He provides a guide for enlightened action. Perhaps he even inaugurates a discipline, combining engineering and strategy – what a task! Engineering is the discipline that consists of designing and building computer systems and, more generally, scientific applications. From here, engineering can be transposed to human affairs, businesses and organizations, including the military – strategy being the art of leading an army – as Jacques Arcade proposes, and it is a step he takes without hesitation and with conviction. Is it possible to compare technical-industrial systems and human organizations? Certainly, in both cases, one must have foresight and knowledge in order to be able to do so. In human organizations, however, there is no mechanism, everything is in motion and is even unexpected and unpredictable. Thus, it is necessary to renew anticipation while acting in an environment that is itself evolving. It is this complexity that strategic engineering sets out to tackle.

On the one hand, in a socioeconomic management book named *L'ingénierie stratégique du roseau*, Henri Savall and Véronique Zardet defined strategic engineering as the set of concepts, methods, tools, and techniques for the control of human processes, making it possible to improve the relevance, efficiency, and effectiveness of an organization's strategy (Savall and Zardet 2005). They positioned strategic engineering between a company's strategic vision and implementation in order to develop the competitiveness of companies and organizations. The issues facing strategic engineering are therefore at the heart of management.

Jacques Arcade, on the other hand, defines strategic engineering as "the art and manner of producing that which is "strategic", i.e., the capacity for anticipation and action during the "journey"". Strategic engineering can be summed up by three phases: discerning, arbitrating and intervening. The goal here is to fill the gap between foresight and strategic action, that is, to work on the path that lies between the vision for the future and the present action, keeping in mind that the situation is constantly changing. Here, the goal is the path. The project is the path. How can we adapt our strategy, maintain our anticipation, evolve intelligently, while walking and acting within a complex context? And yet, Nietzsche said that we can only really think by walking. And as it says in the Bible: "I am the way, the truth, and the life [...]." From this it is already clear: the primacy of the path, guided by vision, opens the way to life. And it is indeed the "vitality" of organizations that Jacques Arcade proposes as the goal of strategic engineering.

Jacques Arcade broadens the field of strategic engineering to all types of organizations, including nations, and provides a theoretical and methodological corpus to renew decision-making and guide action in institutions. Naturally, executive managers form the targeted readership of this book. Here, he puts forward a "toolbox" that is a coherent set of methods, concepts such as "guidance" rather than piloting, as well as presenting computer tools and software to assist strategic engineering. But in addition to managers, this book is also aimed at political actors, administrative leaders, and military officials.

Arcade also proposes to go beyond the classical approaches to strategic foresight, whose linear, apparently rational, path would go from foresight to action, via decisions. Inspired by Lucien Sfez's criticism of decision making, which is ternary and rational in equal measure, giving the subject the impression that they "decide" on their own, when in fact this is a multirational process in which different forms of logic collide, Arcade insists on the process and its evolutionary nature. He shows that strategic engineering is a process of anticipation and systematic action at several levels. He draws inspiration from the fractal theory of the mathematician Mendelbrot to adopt a systemic vision, in feedback loops, where finally the institution is compared to a living being, evolving in a dynamic ecosystem. This leads to an increase in the complexity of this book, following a helical movement inspired by the form of a strand of DNA; only readers who follow the author's demonstration step by step will be able to grasp its full richness.

To write his book, Jacques Arcade drew from his long experience of assistance and consulting in companies. This allows him to illustrate his work with numerous tables, figures and examples. He has also built sophisticated software tools that help in the implementation of strategic engineering and that legitimize an approach that is not only academic and heuristic, but also intended to be put into practice in the management of an organization.

In times of crisis, when the "cult of urgency" always carries the day, is it necessary to recall the need for foresight and anticipation? This is not a question of a foresight process that is enclosed in reports or scenarios, but of an approach that draws paths and modifies the representations of actors, and consequently, informs strategic actions. The challenge is to shed light on the medium-to long-term outlooks of organizations, companies, and even countries. The serious global crisis that arose in 2008 left in its wake a blind short termism, paving the way for in-depth reflection on the regulation and sustainability of choices and strategies. In "liquid" societies and organizations, as Zygmunt Bauman says, one must multiply the "grab handles" we can hold onto in order to orient ourselves and act. Arcade's book comes at the right time to shed light on reflection and action and offers some major "handles" to anyone willing to take the time to read it.

Pierre MUSSO
Professor at University of Rennes 2
and at Télécom Paris
December 2020

Introduction

The world is being shaken by economic and financial, geopolitical, socio-cultural, ethno-religious, health and epidemic-related, ecological, and climatic upheavals that foreshadow or bring about the advent of a new era for which we are insufficiently prepared. This new era, which we are insidiously making ever more inevitable, both through the persistence of our behaviors and through our relative apathy to the consequences, proves to be all the more drastic as it manifests itself in unexpected and unique ways on multiple scales of space and time. How could we, at our respective levels of action, contribute to confronting this strange adversity and its various implications in terms of uncertainty, perplexity, anxiety, and even disarray, against the backdrop of planetary crises? What methodological approaches and what primary skill sets can we see ourselves implementing to intervene rationally, taking advantage of such a situation, to overcome the "climate of downturn" and to open up new horizons for the long term?

The strategic engineering approach, the subject of this book, is an original contribution to meeting such a challenge. The transformative action that results from this approach proves to be well-reasoned and innovative, as shown by the organizational innovation it generates. Moreover, this action is cardinal, that is, decisive and archetypal, because of its pivotal nature and the agility that results from it. Moreover, this action is designed to be both engaging – right from the design phase – exhilarating and, throughout its implementation, even edifying – thanks to the appropriation that it generates and the virtuous dimension that it conceals and infuses.

Strategic engineering revolves around an integrated methodology for the creation and processing of strategic interventions, with the goal of assisting in carrying out the appropriate design, contextualized evaluation, and deployment of decisive interventions of a structuring nature.

The initial concept of strategic engineering is the result of a long period of development beginning in the 1980s, based on a systemic representation approach of the context of the intervention and the monitoring and evaluation of the situational strategic action. This approach was completed, at the end of the 1990s, in the form of a holistic and unifying reference framework or metamodel. Since then, various aspects of this framework have regularly been the subject of university master's courses, training seminars/workshops and *in situ* work sites. In addition, strategic engineering has given rise to the development of a demonstrator integrating the specialized tools succinctly presented in the book. This period of maturation of the strategic engineering approach has benefited from diversified experiences, both at the national and regional levels and in the framework of international organizations. This experience has been extended in key sectors of industry, advanced technologies, and services, including consulting, both within the framework of international structures and within the framework of international organizations and privately owned companies, as well as for public, civil and military administrations and organizations.

Strategic engineering is based on a conceptual, instrumental, and process-based interconnection between foresight exploration and strategic intervention. It seeks to combine conceptual relevance and operational excellence. It is in line with the fractal geometry movement, according to which many natural or ideal objects present, at each scale of observation, a complex, fragmented and irregular structure, which proves to be, following a principle of "self-similarity", analogous to that of the whole. Strategic engineering also draws inspiration from other aspects of complexity sciences, in particular by using the concept of the successive embrace or accretion of nested frames at different scales. This mechanism of abstraction at multiple levels of "meta-" order also corresponds to a spiral elevation, or rather a 3D vortex, leading to a phenomenon of dynamic "coalescence". This phenomenon induces a form of two-way emergence; this is to say, at once ascending, bringing together certain properties resulting from the interactions of the lower level, in the form of constraints, and descending, integrating certain properties resulting from the interactions of the lower level in the form of constraints, causing the lower levels to conform to the requirements of the higher level, which these have helped to generate. Strategic engineering draws inspiration from this principle of double emergence to map the context of strategic action. It is also enriched by methods borrowed from other disciplines (the sociology of organizations, policy analysis, integrative biology, physiology of action, object-oriented approach, etc.). It further uses various techniques of organization and decision support (i.e. a goal tree, table of actors' strategies, multicriteria selection, game and graph theories, convex optimization, a project logical framework, etc.). It does not claim to reinvent existing tools but works to bring them together, so as to react to them in a judicious and appropriate way, to complete them and to concoct a judicious and relatively new synthesis.

Without denying the pedagogical nature of a sequential or synoptic approach, strategic engineering has the advantage of being structured like a metamodel and thus can approach and solve questions relating to the different stages of the project cycle (anticipation, design, decision, implementation, monitoring of execution, and feedback) along the way, thanks to the flexibility of its systemic representation mode at several levels.

The underlying principle of this method, which stems from its praxeological dimension, implies that the evaluation of the overall performance of the intervention results from an assessment based on three complementary macrocriteria or viewpoints: finalism (appropriateness to the strategic frame of reference), pluralism with regard to the future (validity with regard to possible futures or "futurables"), and pragmatism (congruence with the intervention environment and context).

These criteria refer respectively to the main subsystems that make up the overall framework or "strategic landscape" of the intervention: the strategic frame of reference, the foresight scenarios, and the surrounding context. Thus, the hierarchical breakdown of the intervention's framework makes it possible to move systematically from the structure of nested frames to a tree-shaped structure of criteria describing the situation in its context; then to a "cascading" weighting of these criteria, in order to evaluate the strategic intervention in light of these criteria. Moreover, this breakdown can be carried out at different levels of aggregation. Strategic engineering integrates a process, characterized by the art of dialectcally and dynamically binding together the situational context, the vision, and the systemic relations between the participants, up to the limits of their space of initiatives. Thus, the "strategic", as a substantiated adjective embodying the essence of the underlying process of dissemination of decision and action, supplants and transcends the notion of strategy *stricto sensu,* once immersed in its situational context. This suggests the possibility of accessing a more subtle form of "physiology of strategy".

Strategic engineering favors the notion of adequacy over that of deliberate orientation or of a desired end state; that of guidance (autonomous, intuitive, and adaptive) over that of piloting, which has a mechanistic connotation; and that of effectiveness over that of efficiency, which is both more normative and restrictive. Beyond the traditional approaches in this area, strategic engineering is based on a fluid and harmoniously disseminated operational paradigm, so as to fit within the global context or strategic landscape of the action. Strategic engineering is a methodological and logical device for conceptualization and perspective setting, including some advanced tools of an unprecedented nature. As such, it constitutes a much-needed tool for anticipation, both meticulous and comprehensive. An original "tooled approach" process, combining optimization and post-optimization, allows anticipating and remixing any recommended solution in the best way possible, on the basis of the numerous parameters of the decision-making situation. Each solution profile can then be visualized through a triptych of dashboards allowing visualization and interpretation of most of the aspects and characteristics of this situation.

Strategic engineering is rooted in the mysteries of sociopolitical activity and its situational context. As such, we can no longer be nostalgic observers of an illusory world, so often typified by the glory days of the post-war era, in continuous expansion. From now on, we must mobilize ourselves in concert to try to contain, and even overcome, the alarming specter of the chaotic, ambiguous, and unpredictable world that is emerging at the dawn of the 21st century.

The strategic engineering approach proposed in this book is deployed in three phases so as to inform transformative action. These phases correspond, respectively, to the three parts of the book:

– observing and discerning: contextualizing the situation and recommending a consistent set of options (Part 1);

– judging and arbitrating: evaluating and remixing recommendations to decide on a successful intervention (Part 2);

– acting and intervening: judiciously guiding actions to generate virtuous dynamism (Part 3).

Each phase consists of three stages, which are the subjects of the nine chapters of the book. These steps can be reconfigured and rearranged, if necessary, by each reader on the basis of his or her expectations or interests.

This book contains some examples of concrete cases in which the author contributed to the execution. They are concise, circumscribed, and far enough in the past to emphasize their educational value and illustrative nature.

PART 1

Observing and Discerning: Contextualizing the Situation and Recommending a Consistent Set of Options

1

Understanding the Strategic Landscape

It is not possible to grasp and resolve a critical or difficult situation without first trying to put it into context, and in particular, within the immensely complex interplay of its main participants. Moreover, the strategist must also take into consideration the lines of force that tend to overdetermine the situation, as well as the way in which it is approached and perceived. All of these elements are part of what we will call the "strategic reference".

From this point of view, it is necessary to adopt a global and synoptic representation of the situation as well as of the strategic landscape – that is to say, of all the aspects involved – in which they take place. On the other hand, this frame of reference must also be penetrating and sharp enough to meet the needs of the analysis. Indeed, it must allow a telescope-like access to the characteristics of the different essential components of the situation, including the ability to dig deeper and "zoom in" on the strategic landscape, depending on the circumstances.

It is important to enrich this view with a "phenomenological" dimension, which must be able to integrate the mechanisms by which the components of the context and the strategic interventions would be able to be coupled, or even combined, in order to evolve together and adjust to each other. Finally, this reference grid should also include a dissipative dimension, so as to account for the emergence of richer and more subtle properties. This will be the same, as we move towards more global components, located at high levels of aggregation.

In order to guarantee the set of requirements imposed on this vision, and thus on the approach they infer, we have chosen to articulate the representation of the strategic landscape around the concept of "nested frames".

1.1. Understanding the decision-making situation

First of all, the environment of the decision must be taken into account. Made up of different facets, it must be considered from its "salient points", that is, the critical and decisive elements for the entirety of a problematic situation. Any situation of crisis needs to be mapped in order to provide an intelligible representation of its context and the nature of the system to be studied. The context will be referred to here as the "strategic landscape". By avoiding the trap of an overly exhaustive approach, a structural analysis can help identify the particularities of the situation and clarify the configuration of its main facets. The strategic landscape is made up of three macrocomponents: the strategic frame of reference, the future general environment, and the specific context or theater of intervention. Once the strategic landscape has been established, it must be represented using a figurative approach by means of a succession of nested systemic frameworks (or components).

1.1.1. *Identifying the variables and factors that make up the situation*

The first course of action is to place the limits of the field of reflection or study. For this, the first step is to reach agreement on the question that forms the basis and justifies the reflection. In fact, on the one hand, we cannot expect to provide solutions to false problems, and on the other hand, there is no system that can act as a mode of representation of a reality without a prior intention, even if only implicit. It is therefore important to list all the variables and/or factors, both internal and external, characterizing a system or likely to affect its evolution in the prospective horizon. At this stage, it is important to be as exhaustive as possible, taking care to avoid gray areas in the description of the system or its seeds of change. In addition to the collective reflection and brainstorming meetings, it is advisable to foster and support the establishment of the variables through directive interviews without leadership, especially for particular areas where the group lacks expertise, as well as with personalities with particular knowledge of the presumed actors within the system.

In the second step, the list of variables is then formatted, completing it if necessary, but also grouping or breaking it down, or even eliminating it, in order to arrive at a list with a sufficient level of homogeneity.

After first classifying the variables according to categories that allow for a more precise distinction between internal and external elements of the system, an effort is made to formalize their meaning within a multidisciplinary group created for this purpose.

The formation of this group is carried out both on the basis of the professional responsibilities and the *intuitu personae* skills of these presumed members. This will primarily involve facilitating comprehension using simple headings to avoid any misinterpretation by persons outside the group. For each of the variables, we then add:

– its definition in the perspective adopted by the group;

– a mention of the problematic elements that it raises (highlights) and any indications of its retrospective and possible future evolutionary trends;

– an appreciation of possible trend breaks, whether presumed, desired, or feared.

It is important to mention that any discussion of problems that may arise during the collection of variables may feed into the rest of the process. This phase is fundamental because it is an excellent opportunity within the group to build a common reference to represent and in this way understand the system. Moreover, it encourages the decompartmentalization and cross-fertilization of different points of view. The potential identification of the relationships between the variables that form the core of the structural analysis is substantially improved, which helps to stimulate the appropriation process within the group.

1.1.2. *Detecting the context of the situation through structural analysis*

How, and in what way, can the application of a structural-type analysis help, beyond the construction of the information base necessary for the modeling of the strategic landscape, to reveal context through the identification and assembly of relevant nested components?

1.1.2.1. *Main steps*

In order to answer this question, we will limit ourselves to recalling the main steps of the method known as "structural analysis" (see Box 1.1), in order to focus on the interpretation and visualization of the results obtained.

1) *Identification of the variables and factors making up the system.*

– Review of available information and studies concerning the system.

– Identification of factors of change and key variables.

– Detailed formulation and establishment of the glossary of variables.

2) *Development and processing of the structural analysis matrix.*

– Systematic reconstruction of the direct influence relations between the variables.

– Establishment of the synoptic table of cross-influences corresponding to the structural analysis matrix.

–Computer processing of the structural analysis matrix.

3) *Visualization and interpretation of results.*

– Keys to interpreting the influence/dependency plan.

– Summary and taking the results of the structural analysis further.

Box 1.1. *A three-step analysis*

An illustration of our argument using the case of a "developing country" allows us to reconfigure the cloud of variable points, which constitutes this country-wide system, in the form of nested homogeneous components, or "nested cores". The homogeneity of these components stems both from the semantic proximity of the variables they incorporate, and from the similarity of the role they might play in the dynamic and long-term evolution of this country. This example has the benefit of showing that, at this early stage of the strategic engineering process, it is possible to identify the systemic importance of the variables, and to draw valuable lessons – including strategic lessons – from such an analysis.

1.1.2.2. *Key areas for interpreting the results*

To facilitate its use in this framework, it is important to present this grid through the description of the projection plane of the variable points submitted to the analysis, as well as the corresponding key areas for interpretation. The variables characterizing the system to be studied and its environment can be projected on the graph presented in Figure 1.1 (section 1.1.2.2.1), in which the vertical (y) axis corresponds to the influence and the horizontal (x) axis corresponds to the dependence.

The projection of the cloud of variable points on this graph allows us to make groupings by categories, based on the semantic approximation of the variables concerned and on the proximity of their positioning in the plane. Thus, these categories are differentiated according to the specific role their respective variables are likely to play in the dynamics of the system. The refinement of these categories makes it possible to establish an indicative typology of the variables, which proves to be valuable for the understanding of the strategic landscape.

The next two sections, supported by figures, recall the determination of the main regions of the plan (section 1.1.2.2.1), as well as the indicative location and interpretation of the clusters of homogeneous variables (section 1.1.2.2). Finally, an example of a relevant grouping of variables is presented, in which nested components or nested kernels have been gradually incorporated into clusters (section 1.1.2.3).

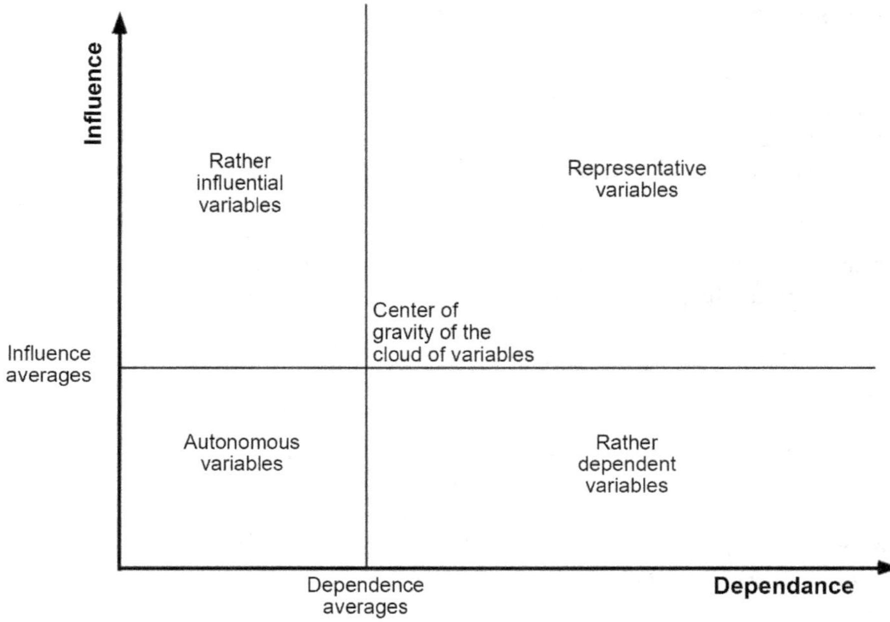

Figure 1.1. *Preliminary interpretation grid for variables in the influence/dependence graph, following four major areas*

1.1.2.2.1. Determination of the main categories of variables: representation by a point cloud

The variables characterizing the system to be studied and its environment can be projected on the influence/dependence plane, that is, a plane where the y axis represents the influence and the x axis the dependency. The repartitioning of the scatterplot variables on this graph – and in particular with respect to the different dials formed around their center of gravity – allows us to determine four main categories of variables.

These categories, which differ according to their specific role, can play a role in the dynamics of the system.

These categories are:

– *Rather influential variables*: these variables are both highly influential and not very dependent. These variables, which are located in the upper left quadrant of the graph, provide the conditions for the essence of the system.

– *Rather dependent variables or result variables*: these variables, located in the lower right quadrant of the graph, are both low in influence and highly dependent, and are therefore particularly sensitive to changes in influential variables and/or relay variables. They are output variables of the system.

– *Representative variables*: highly influential and highly dependent, these variables, positioned in the upper right quadrant of the graph, are by their nature factors of instability, in the sense that any action on them leads – provided certain conditions in other influential variables are met – to repercussions on the other variables. However, these impacts may have a boomerang effect that either amplifies or defuses the initial impulse.

– *Autonomous variables*: these have little influence and are not very dependent. These variables, located in the lower left quadrant of the graph, appear relatively disconnected from the system, in the sense that they do not allow a large-scale progression to be carried out within the system, nor make it possible to take advantage of such a progression.

The configuration of the entire scatterplot of variables allows us to appreciate the *degree of determination or predictability of the system*.

The farther the scatterplot stretches along the horizontal and vertical axes, the more it can be considered to be relatively predetermined and stable: the response of the system to a given impulse of the determining variables is significant and relatively predictable.

On the other hand, when this cloud extends around the first bisector, the system can be considered as not very predetermined (unstable), especially as the points approach the northeast end of this diagonal axis. These variable points, which are characterized by both high influence and dependence, will play an ambiguous role. Acting sometimes as drivers, sometimes inhibitors of the dynamics, they constitute factors of uncertainty to anticipate the transformation of the situation, which also evolves according to the variables judged to be determining.

It should be pointed out that this notion of determination or structural stability does not mean that the system will not evolve. With all other things being equal, the impacts of the variables influencing the system's dynamics will be unpredictable; in particular, there will be very few boomerang effects.

1.1.2.2.2. Location and interpretation of clusters

The projection of the representative points of the different variables onto the influence/dependence graph has led to relevant groupings of these points into 11 "packages" or homogeneous clusters, so as to enrich the interpretation grid. The provisions in the influence/dependence graph, as well as the respective categorizations of these different subgroups, are shown in Figure 1.2.

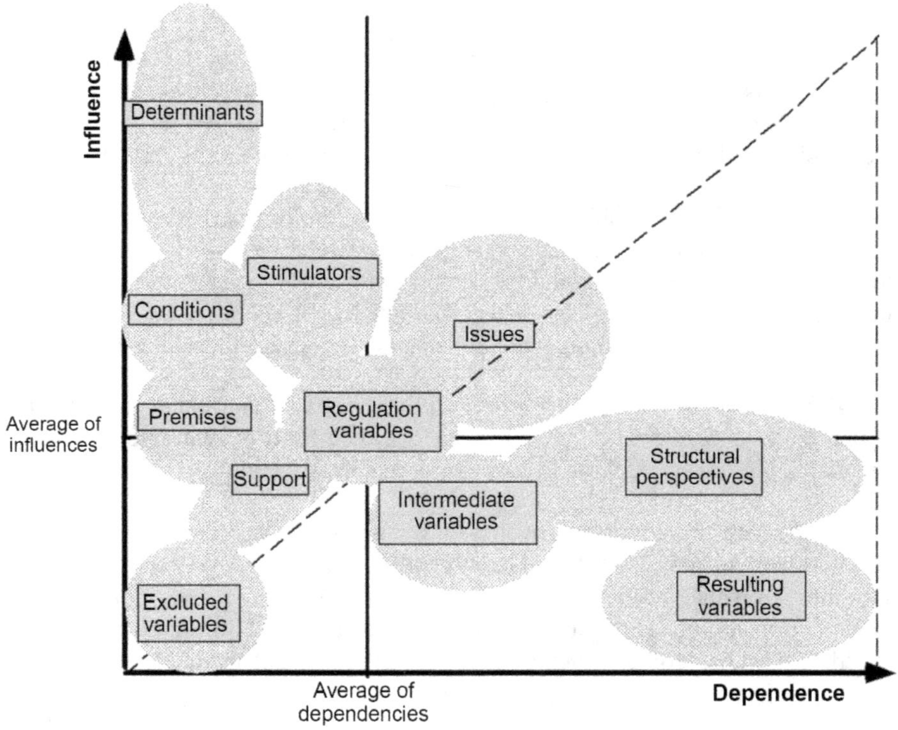

Figure 1.2. *Refined interpretation grid of variables in the influence/dependence graph on the basis of approximately ten sectors*

The indicative nature of these packages of variables/factors is intended to facilitate the understanding of the results of the structural analysis. From this point of view, these packets are successively reviewed below; they suggest different interpretations of the effective role that the variables contained within them could play in the evolutionary dynamics of the system under study.

– *Determinants*: these variables and factors are mostly located along the upper-left axis formed with respect to the system's center of gravity. They can act, depending on the degree of mastery and control available, either as factors of inertia or as key factors of system mobility. These factors are also considered as input variables for the system and most often include the environmental, or rather, exogenous, variables that weigh heavily on the system, while being relatively uncontrollable. They will instead intervene as factors of external influence and/or vulnerability of the system towards the outside.

– *Stimulators*: these are the main variables through which some control over the system can be exercised. Essentially, they have an effect either as factors of modulation of the external influence or the exploitation of exogenous conditions, or as control variables, impulse factors, or triggers of action within the system.

– *Conditions*: these variables are located below the determinants. They are therefore less centered than the determinants and operate relatively further downstream. Because they are generally beyond the control of the actors in the system, they act as contextual or environmental elements which, through their hold over the system, form the condition for their evolution, without being the root cause for such change.

– *Premises*: these variables are a little further back than the "conditions" and can be expectations, which in some cases may be mobilized by the actors, with the potential to play a catalytic role in the evolution of the situation.

– *Challenges*: these may be coveted by the main players, since mastering them can alter their capacity for action. Given their unstable nature, they are a necessary point through which influences and dependencies must pass or even a pivot around which the system rotates. Acting directly on these variables is difficult and delicate. Indeed, the impact of any action in this direction remains uncertain and can sometimes give rise to "boomerang effects". It is therefore advisable to consider these variables as important presuppositions for the systemic evolution of the situation under study. However, attempts to try to control them as issues constitute a major challenge that can only be met through a proactive vision or a way of overcoming them.

– *Structural perspectives*: these are located between the issues and the resulting variables, rather towards the upper end of the southeast quadrant. They can therefore be considered, to a certain extent, as identifiable effects of the evolution of the system, but on which a voluntarist action can be taken to make them evolve in the desired direction, for example, to escape a certain fatality. These are more possible aims of the system as a whole than predetermined or programmatic consequences.

– *Excluded variables*: these are closer to the origin of the axes. Their evolution therefore appears to be relatively autonomous and disconnected from the overall dynamics of the system.

– *Support variables*: although they are relatively autonomous, these tend to operate upstream. The variables concerned can be used as auxiliary action variables or as a point of application for possible accompanying measures.

– *Mediation and regulation variables*: this type of variable deserves to be cited less for its intrinsic definition than due to the originality of its location with respect to the other types presented above. These are variables that are located around, or rather above, the center of gravity of the system. They may come into play as secondary levers or stimulators, as weak objectives, or as auxiliary stakes.

– *Resulting variables*: these are highly dependent. They therefore result from the overall evolution of the system and are therefore excellent synthetic indicators of its overall state. The corollary of this property is that they require numerous endogenous conditions to be set in motion in the desired direction. These variables can therefore only be activated in a significant way by *ad hoc* measures or actions.

– *Intermediate outcomes*: these variables are both more influential and less dependent than the resulting variables. They can be viewed as intermediate outcome opportunities where there is little scope for rebounding to structural prospects and/or influencing overall system outcomes.

1.1.2.3. *Mapping "in the abyss": the case of a system for a developing country*

The elements presented in this section for essentially educational purposes are the result of a concrete experience, carried out with the support of the United Nations, within the framework of a prospective study of a developing country over a 30-year period. The system studied consisted of about 70 variables. These variables had been previously classified *a priori*, according to the categories commonly accepted in national planning. These variables were then grouped at two successive levels, grouped in a way that could be typified by the form of a Russian doll, as shown in Figures 1.3, 1.4 and 1.5.

At the end of the first stage, less than 20 components were identified. These were ultimately grouped into seven macrocomponents, identified and named accordingly, as shown in Figure 1.5.

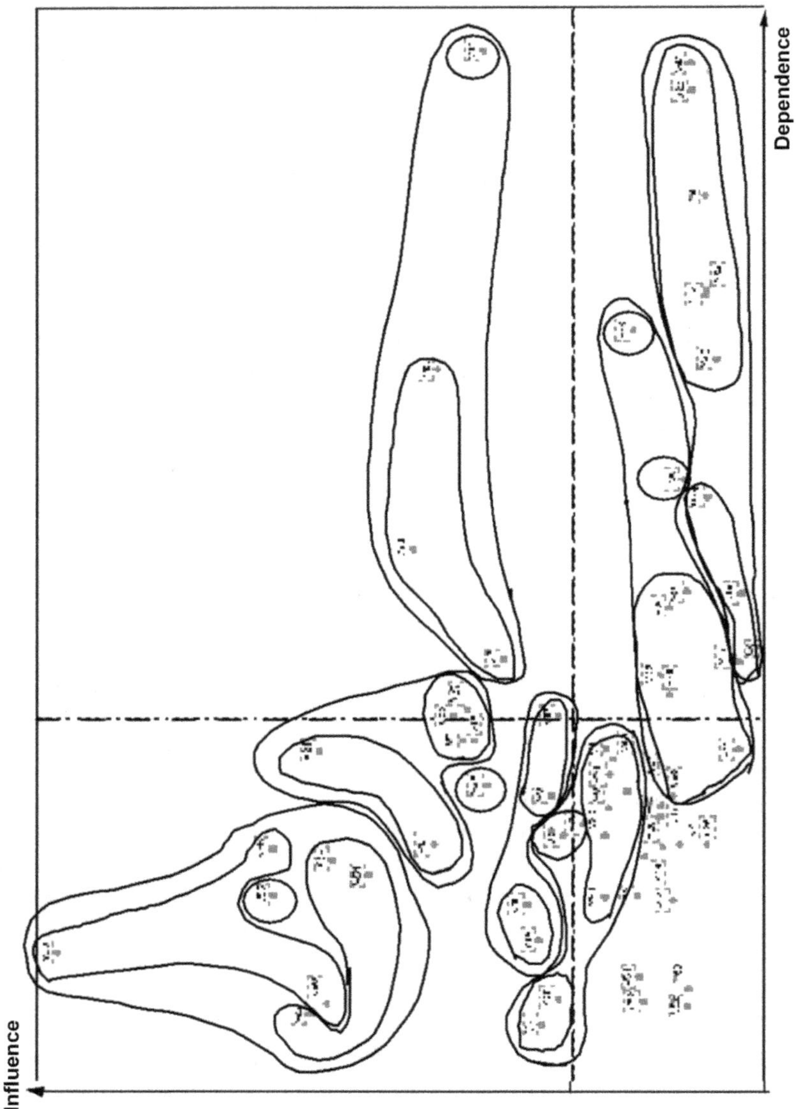

Figure 1.3. *Structural analysis: positioning of system variables*

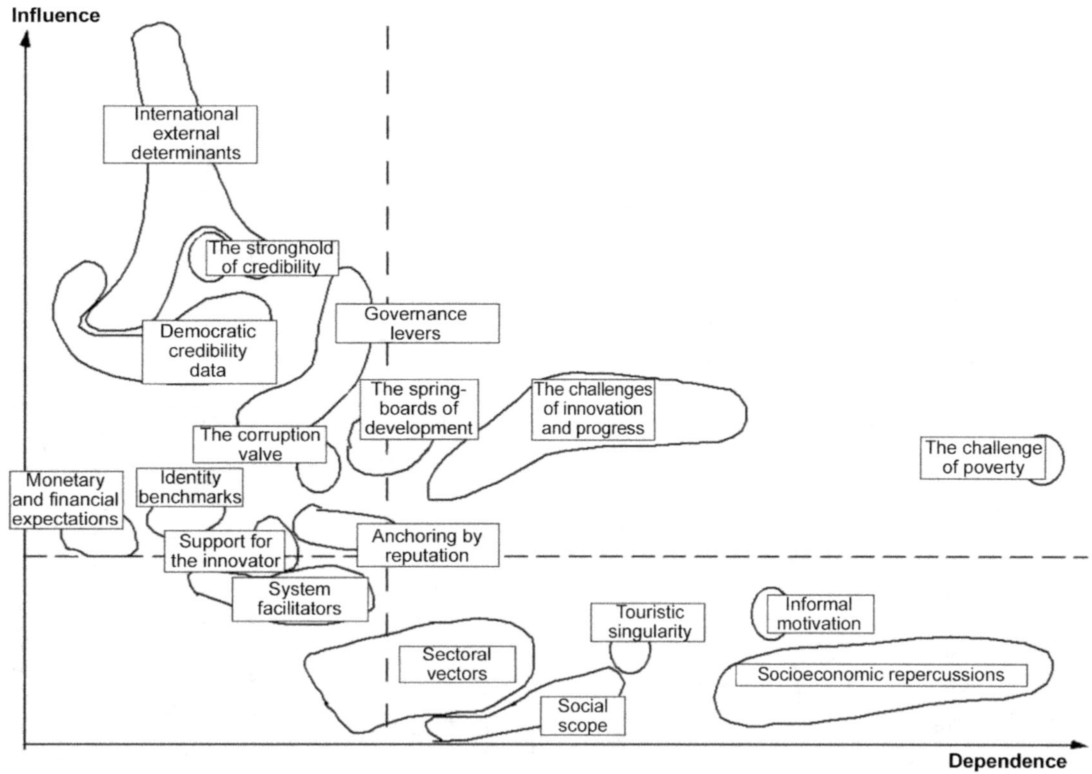

Figure 1.4. *Structural analysis: specification of packages of variables*

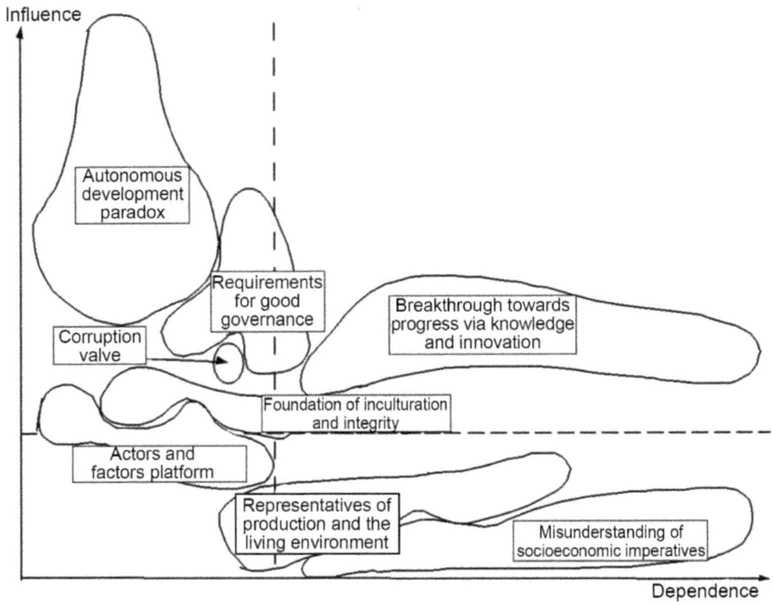

Figure 1.5. *Structural analysis: general configuration of the system*

1.1.2.4. *Key messages from the analysis*

As an indication, the key messages resulting from the structural analysis could lead to the preliminary formulation of structuring recommendations, in the sense that they are more at the level of the landscape and the strategic frame of reference than interventions or actions. In the illustrative case of the developing country (Mali), these points of recommendation can be sketched out for illustrative purposes, as shown in Box 1.2.

Key recommendations from the case of a developing country

– Forge, within the current geopolitical context of the subregion, regional integration as the country's "spearhead" on the international scene.

– Aim to be progressively more capable of putting economic and financial imperatives into perspective through making good governance a best practice.

– Enable the country to take ownership of new technologies in order to take a sustainable place in the innovation of multimodal services within the subregion.

– Strive to interconnect and reconcile, on the one hand, "breakthroughs towards progress" – via the management of knowledge and skills – and, on the other hand, "poverty reduction".

– Rebalance the importance and content of the mainly sector-based approach imposed by donors in favor of enhancing national identity and consolidating cultural integrity.

Box 1.2. *Summary of key recommendations from the case of a developing country*

This will be a continuation of the strategic engineering process begun at this stage of the analysis, as follows:

– amending and validating these recommendations by *re-appropriating* them in an initial step;

– enriching them a and making them available in concrete orientations, in a second step;

– contrasting them with the different exploratory scenarios, once prepared.

1.2. Representing the context of the action

The founding concept of strategic engineering is based on an approach centered on multilevel systems, which gives rise to the notion of a "system of systems". Such a system infers a model of the strategic landscape known as a "nested frame", for which some of the principles are summarized in Box 1.3.

The term "frame" was introduced by Minsky (1997) and can be defined as the representation of a context. According to this author, "using the language of frames, it can be said that the mental representation of a situation implies the use of frames embedded within each other. Each of the ingredients of a situation has its own frame... The theory of knowledge representation in the form of frames is based on the idea that the world is made up of almost closed sub-systems, each of which can be used as a context without being too overwhelming or creating too much upheaval in the process. Each frame thus has a set of expectations of its predecessors, in addition to the demands it imposes on its subsequent frames".

Furthermore, Mandelbrot (2010) sheds light on this concept, which evokes the fractal geometry he invented, by declaring that: "Fractal objects do not (only) invoke a descending cascade, namely the fragmentation of very large, very diffuse masses into smaller and smaller pieces, but (also) an ascending cascade, namely the clumping of very diverse dust into larger and larger pieces. The problem is very similar to that of waterfalls in turbulence theory. However, the most recent results in this field suggest that the two kinds of cascades coexist: thus, there is hope that disputes between proponents of coagulation and fragmentation will be resolved".

Box 1.3. *Nested frames and cascading contexts (Minsky 1997)*

This modeling involves considering this landscape as a system, then breaking it down into three interacting subsystems: the strategic reference frame, the general environment scenarios, and the general context of intervention, the latter of which includes in particular the interplay of actors.

1.2.1. *Addressing the situational context through nested frames*

In accordance with the concept of nested frames, the subsystems of the strategic landscape are considered as components, broken down to a finer scale to capture details and provide a more in-depth the analysis. For example, the *future general environment*, constructed through morphological analysis, consists of a range of global prospective scenarios. These are, in turn, conceived as alternative systems of thematic scenarios, which result from coherent conjugations of hypotheses (or elementary scenarios) about the future.

To implement this approach, two complementary mechanisms are used to describe these complex systems:

– the component tree (or subsystem tree), which reflects the vertical hierarchy of the levels of analysis and therefore of the associated contexts;

– the horizontal interaction mapping, which describes the components within each level.

In this approach, each component of the system at a given level serves as a context for the components it encompasses, as it provides the conditions for its structure and the connotations of its content. Thus, each component of the strategic landscape is immersed or "subsumed" within multiple layers of increasingly broader contexts. While preserving its characteristics and singularities, this component proves to be subordinate to the essential properties of its contextual framework, which it in turn contributes to making specific and concrete. It allows for "contextualizing" the downstream components, which it incorporates. For each component, this process of cascading presentation can be extended, up to the level of detail deemed relevant. For example, the hypotheses on the future, forming the thematic scenarios often considered elementary, could be expressed where appropriate as microscenarios, built on the basis of more detailed hypotheses.

This level of "resolution" is therefore contingent on the situation being studied and is consistent with the "criticality" of the information associated with each component. In addition, each of these components "inherits" a configuration, in the form of a matrix of interactions whose intelligence mode and keys for interpretation present a certain analogy at the different levels of analysis. This characteristic property of the so-called "fractal" geometry or "holomorphic" structures presided over the modeling of the strategic landscape.

Grasping the landscape in this way generates significant savings in organizing the information required and systematizing the form of reasoning used. Indeed, the operative quality of the representation of a strategic landscape depends, among other things, on its consistency and congruence to the situation; that is to say, on the way in which the characteristics of its components at different scales resonate.

This property has the potential to enrich the description and reinforce the relevance of the strategic landscape. Indeed, any attempt – including large-scale attempts – to describe or modify a component that is at odds with the rest of the strategic landscape runs the risk of collapsing under the weight of its contextual strata. In the same way, initiatives – even of limited importance – that may increase or exploit the similarities between the scales (of the context and/or content) of the components create a synergistic and resonant effect of a considerable, even unmeasured magnitude, similar to the levels of energy released in certain natural phenomena (storms, tornadoes, or earthquakes).

The nested frame method thus represents a contextualized approach to the strategic landscape, which we will describe as "hyper-contextual". This method makes implicit use of the principle of subsidiarity with regard to the degree of analysis of each component, allowing for the optimization of useful information to describe and understand the strategic landscape.

In order to facilitate the implementation of the nested frame or component approach, it is sometimes useful to use so-called "structural" methods, according to the instructions given in section 1.1.2.3.

1.2.2. *Conceptualizing the strategic landscape as a system with three macrocomponents*

The strategic landscape is the context in which the intervention will take place. Defining this context makes it possible to characterize the intervention and determine how suitable it is to be implemented. It also becomes a system for evaluation prior to and subsequent to any action undertaken and/or executed. In effect, it means constructing a lively backdrop for the intervention.

The strategic landscape has three components which form the possible approaches to action and the areas to be taken into account when assessing it:

1) *The strategic reference frame* is structuring and relatively unchanging in relation to the interventions. It consists of the major orientations, the strategic vision, the essential vocation, and the fundamental capacities (or key skill sets). The strategic frame of reference includes the "driving force" that will inspire the decision-making process.

2) *The general future environment (or general environment scenarios)* is described in the form of contrasting exploratory or prospective scenarios and allows the long-term validity of the interventions to be assessed with respect to possible futures.

3) *The general context of the intervention* makes it possible to evaluate the capacity to integrate the interventions in their concrete and immediate context. It is composed of the politico-strategic context understood in terms of the interplay of actors and the legal/regulatory, technical/economic, and other such contexts.

It is possible to associate a tree-shaped hierarchical structure with each of the three components of the strategic landscape mentioned above. This structure can be extended as much as the understanding of the situation requires.

The reconstitution of these various tree structures upstream is done in a differentiated way according to the space considered. This differentiation ensures the approach is effective.

1.2.2.1. *The strategic reference or major orientations*

The strategic reference (SR) has two systemic components: first, the vision, which consists of its primary motivation and dictates its overall action, and second, the fundamental capacity, which confers a distinctive advantage over partners, as well as a reasoned flexibility to play with uncertainties.

The method recommended for the strategic reference consists of laying out, at different scales, the global vision in terms of goals as well as the major long-term political and strategic orientations.

It is thus a question of establishing them in terms of objectives. These steps allow us to develop the tree structure and then the nested frames. This work requires taking into account the pre-existing elements of the problematic situation, requiring strategic intervention. These elements need to be clarified, collected, and synthesized in a hierarchical manner in order to make them more readable and to use them to design and evaluate strategic interventions.

1.2.2.2. *The vision*

The vision to be assumed within the framework of the strategic reference of a given intervention must be at a more global and fundamental level than that of the proposed purposes. As such, the vision constitutes the ultimate level of primordial ambition.

In the absence of more precise strategic alignments and as an indication, we can determine the elements of pre-existing orientations, which are the result of a limited consultation process. However, it is advisable to keep a certain distance from these orientations, which are ultimately formal, in order to complete or revisit them, or even to reinvent them through broader reflection. The goal of this is to reshape the priorities in a less generic and more salient way. This is the price paid in order for the criteria derived from the vision to make it possible to settle on certain strategic interventions during the evaluation.

The vision is that of actors, as they often formulate it themselves. These actors must also specify the place and the role they intend to play to ensure this vision is implemented. The vision cannot be limited either to choose a course or to determine the means to achieve it. It is both more global and more mobilizing.

The challenge for the main actor is to tighten its hold on factors that it is not – as a first approximation – supposed to control. This is a *proactive* attitude, in the sense that it proves to be in the order of strategic achievement: opening up a new space for freedom and creating the favorable conditions that accompany it.

The proactive vision is based on crucial issues, first-order trade-offs and decisive alternatives in terms of the major challenges to be met. It can also take the form of a "driving force" or a "primordial impulse" that can serve as a reference or "impulse code". In particular, this force helps in decisions on decisive trade-offs and in winning support in order to develop, choose, and implement options for strategic action. The notion of a driving force provides a permanent framework for understanding the basis, relevance, and "functioning" of the strategic vision or purpose.

At a minimum, the vision should include an outline of the broad ambitions on which it is based, as well as the obstacles and challenges that must be overcome in order to achieve it.

The criteria for the relevance to the vision are expressed as a coherent combination of goals, fundamental principles, and orientations. These can also be visualized in a tree-like structure. In order to formulate this tree structure, it is important to first clarify the strategic alignments, the guiding principles, and the priorities. The second step is to bring them about to clarify their articulation. The more operational principles at the lower level serve as points of application for possible policy interventions and are comprised of the criteria or partial viewpoints by which the relevance of these interventions will be assessed. Coefficients should then be given to weight the different facets of the vision (goals and fundamental principles), in order to facilitate, if necessary, the calculation of the overall total score for each intervention.

A "cascade" procedure of sticky consensual weighting will be used for the tree structure of the vision. Thus, this type of tree structure offers, in addition to the evaluation of the degree of relevance of the recommended intervention, the possibility of exploring different angles of attack to reach the vision.

In addition, the vision offers a system of purposes and principles that constitute the main components of the matrix, whose interactions must be understood and made explicit, in terms of synergies and possible contradictions.

1.2.2.3. *Core capabilities and competencies*

In addition to the vision, the strategic repository includes the fundamental capabilities or key competencies that the actor/organization possesses or wishes to develop. The conceptual basis for the notions of core capabilities or key competencies refers to the concept of "core business", as recalled in Box 1.4.

The core business is a critical combination of distinctive skills, abilities, and aptitudes (including process execution). This "core business" provides the company with major sustainable competitive advantages and underpins superior profitability and leadership in a product or service category.

A core business must first and foremost generate a major benefit for customers, by offering them functionalities that are perceived as an undeniable advantage. It must also give the company a unique competitive advantage, enabling it to stand out from its competitors to the point of being almost inimitable. The core business must also result from the integration, or rather the particular composition, of several elementary skill sets or technologies that give it its unique character (for example, one of Canon's key abilities is the integration of know-how in microelectronics, optics, and electronic imaging). It must provide access to numerous markets, playing a pivotal role in product diversification and in conquering new competitive spaces (for example, Honda's expertise as an engine manufacturer has enabled it to succeed in markets as disparate as cars, motorcycles, lawnmowers, and generators). Among the lesser known key skills are foresight, anticipation, and customer interface.

In an attempt to define core occupations by exclusion, it can be considered that they are not limited to simple categories in the accounting sense of the term and do not tend to blend together. Rather, true key competencies tend to be refined and strengthened over time. Moreover, they are not limited to the critical factors of success alone and are often not identical to the same set of assets that created past competitive success. They are not a disguised way of describing vertical integration but rather relate to the "virtual integration" of networked enterprises.

Identifying the core businesses in detail requires a significant but indispensable effort, which is more likely to be measured in years or months rather than weeks. The difficulty lies in choosing the right level of aggregation; for example, for Federal Express, should this be considered to be logistics or its various components such as package tracking? This is why we sometimes try to prioritize the different skill levels, which has the benefit of being visualized at different levels of precision.

The methodology for identifying key competencies consists of listing the company's primary basic skills and assessing the degree of mastery of each of them by the company. Can the bearers of this competency in the company be identified? The competencies should be compared with such competencies in other organizations and not only with immediate competitors.

It also consists of verifying the company's effective capacity to combine these skills by examining, for example, certain products which, in the first analysis, are based on skills closely related to those of the company. The establishment of a logical description matrix cross-referencing the different areas of know-how between them is sometimes used for this purpose.

Finally, the methodology consists of highlighting the groups of know-how used in combination.

Developing key skills is a weighty and almost irreversible investment; thus it is necessary for management to make a strong commitment to ensure that efforts are constant in order to become a leader in offering functionality that is highly valued by the customer (for example, Sony has affirmed its desire to be able to offer pocket-sized items; hence the invention of the Walkman, the CD and portable audio player, and the pocket TV). Finally, in order to develop a business's key skills, it is necessary to:

– acquire knowledge;

– integrate it with existing knowledge;

– store it and facilitate access to it.

Box 1.4. *The concept of the core business or key skills of a company (Hamel and Prahalad 1999)*

When the environment proves too chaotic and turbulent, and when, for example, the architectures of the force system recommended in a crisis are hampered by technological obstacles, it is important to take thinking to a more global and anticipatory level, in terms of the future abilities required to remove such obstacles. This attitude consists in seeking, at a higher level of concern, elements that can be grasped in order to concentrate efforts over the long term (acquisition of new technologies, development of industrial partnerships, organizational innovation, etc.). This helps to reduce complexity and dissipate the surrounding complications by releasing pressure, or even the vice, on a "piecemeal" basis.

In this way, the core capacities constitute a "foundation of stability" but also a "redeployment pivot" that can be built upon through significant investments, both tangible and intangible, that have the potential to guarantee the institution's major distinctive advantages, durability, and flexibility in a sustainable manner.

A key capability appears as a *critical* combination of know-how developed according to different major skill areas, such as the area of technology, or more precisely, force systems. Each of these key abilities – which, by its very essence, is composite – is thus constructed as a distinctive profile of know-how on specific fields of competences duly selected according to the *method of morphological analysis*.

Therefore, a core capability, through analogy with the method used for scenarios, no longer has to be represented as a simple combination, but can be represented as a subsystem of key competencies and distinctive know-how, resulting in the specification of a component or subsystem of the strategic repository.

1.2.2.4. *The general future environment or exploratory scenarios*

Concerning the general future environment, the recommended approach is that of scenarios. As a truly proven method, it calls for complete development, which is the subject of Chapter 2.

The description of future environments in the form of contrasting scenarios, as images that are at the same time plausible, relevant, and coherent, is a necessary step to scrutinize the future of the strategic landscape. These scenarios, which are becoming more and more indispensable for decision-making in a complex and uncertain universe, can be used in two complementary ways:

– as a source of inspiration for strategic interventions, thanks to their evocative and metaphorical character with regard to anticipated situations;

– as macrocriteria against which to assess the relevance and long-term validity of the planned interventions, whether or not they are derived from the scenarios.

Within the framework of politico-strategic decisions, it is possible to use the available emblematic scenarios by default (e.g. in the military field, scenarios involving the use of force). These are particularly valuable for describing alternative categories of situations.

1.2.2.5. *The general context of the intervention, including the interplay of actors*

The approach recommended for the general context of the intervention is based on an exercise of successively regrouping and encompassing the "relevant actors" in relation to the situation. We define relevant actors as those who:

– demonstrate homogeneous behavior in the situation considering their affiliation and community of interest;

– are part of the situation and/or are likely to significantly affect its evolution.

There is a tendency to privilege the observation of conflict relations by focusing attention on actors considered either as enemies or allies. However, the increasing complexity of situations requires that we consider all the mutual influences of the various actors that could lead to disruptions or uncertainties. The future general environment is also conditioned by the attitudes and behaviors of the different actors towards each other. Indeed, not only do each of these actors have their own interests and – whether they like it or not – pursue their projects, but they are also motivated to try to influence the other actors in order to achieve their own objectives. This is the case for civil society, for example, which, through the weight it carries with public opinion, can also have an influence on political decision-makers, leading them to reconsider certain strategic decisions.

It is important to gain an understanding of the mechanisms of the interplay of the actors in order to allow for a better understanding of their respective strategies. This is a question of identifying the issues around which the actors will be inclined to get involved, mobilize, and even surpass themselves. Indeed, "today's friends can be tomorrow's enemies" and vice versa. Beyond questions about the importance of interactions, some relationships between actors are ambivalent and have two dialectically connected facets: one antagonistic or agonal and the other synergistic or cooperative. This is why it would seem appropriate to proceed to a systemic representation of the context through a matrix of the actor's strategies. Clearly, certain configurations of the actors' interplay may appear to generate alliances, while others will be the source of conflicts.

It will then be of interest:

– to appropriate and memorize the elements of intelligence of the circumstances and conditions of the intervention, through the understanding of the "base map" constituted by the political/institutional interplay of the actors;

– to empirically build a systemic representation of the political/institutional context, with the possibility to facilitate dialog around the role and the interplay of the actors.

The "politico-strategic context", also referred to as the "sociopolitical context", is made up of actors who are assumed to behave in a relatively homogeneous and significant way in relation to the question or issue under study. Each of these actors has a certain influence on the evolution of the system (possible reactions, projects, etc.). They also have enough influence to shape the context of the intervention, if necessary. Moreover, they are led to assert their ambitions and pursue differentiated strategies to ensure the viability of their respective missions or projects. The mode of representation adopted consists in visualizing the "politico-strategic context" through a two-entry square table, called the "actor strategy matrix". This framework corresponds to a subsystem of the strategic landscape and is presented in the form of a two-entry table with as many elements (rows and columns) as the number of actors considered (Figure 1.6).

The descriptions of the strategic identity cards of the stakeholders (objectives, interests and concerns) are summarized in the boxes on the diagonal of this matrix. However, a system depends as much on its constituent elements – in this case, the actors – as on their interrelationships. This is why the boxes outside the diagonal are filled in as much as possible. They contain the predispositions, attitudes, behaviors, and possible actions that each actor intends to demonstrate or apply towards each of the actors in order to achieve its own ends. Thus, the box located at the intersection of a row and a column synthesizes the factual elements and/or the tentative influence of the actor line on the actor column to ensure the viability of its projects and the achievement of objectives.

The interplay of actors is defined on the basis of acting entities (physical or legal persons, organizations, institutions, nation-states) which form *nested components of* the subsystem of the general context of the intervention. For example, the actor "France" would include the high command of the Armed Forces, the Ministry of Industry and Finance, and the Ministry of Foreign Affairs. France belongs to a particular coalition in the context of a given crisis. Each component of the actor "France" has a particular perception of the situation and its stakes beyond the official discourse.

These nested components can be visualized through a succession of nested diagonal matrix elements.

Each actor at the ends of the branches is assigned a detailed information sheet containing the concerns and presumed intentions of the actor, their strategy, their means of action that can be put into action, etc.

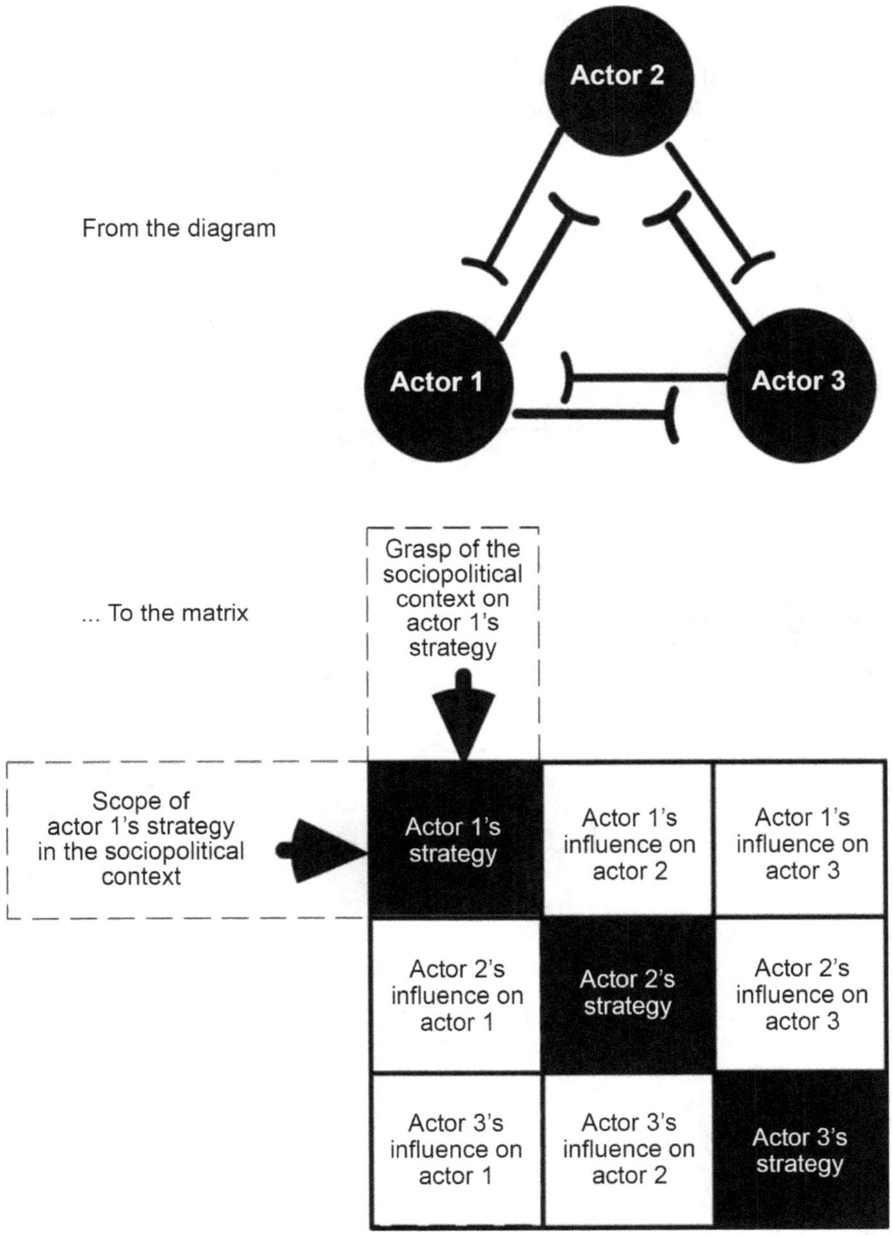

Figure 1.6. *Systemic representations of the interplay of actors*

1.2.3. *Reconstructing the multilevel landscape system*

The context of intervention, including the interplay between the actors, can be synthesized on different scales and in two complementary, relatively equivalent and transposable forms: first, a tree-form representation of the components, and second, a matrix representation of the corresponding diagonal elements..

1.2.3.1. *Tree or matrix representations of components*

These two forms of partial representation of the strategic landscape are illustrated, in the particular context of the interplay of the actors, by Figures 1.7 and 1.8.

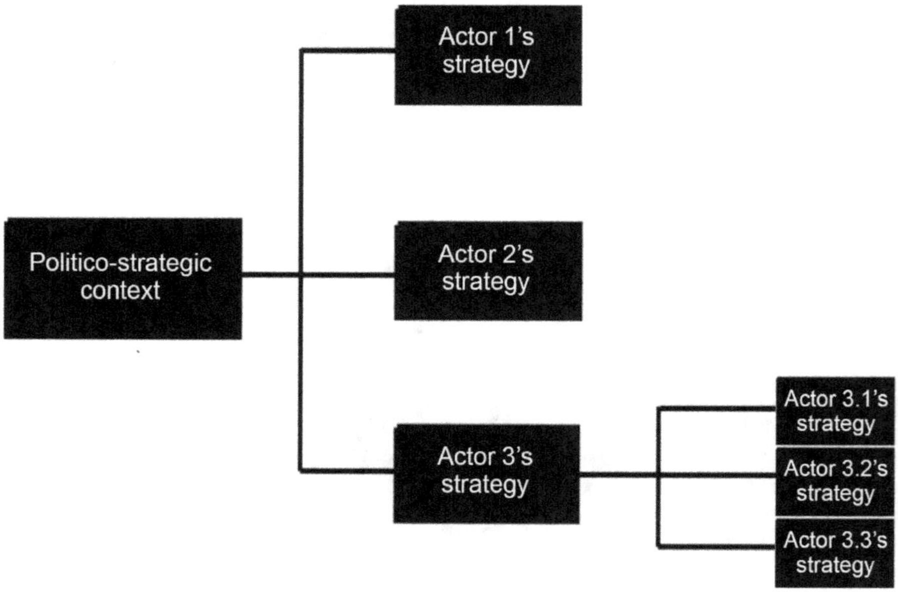

Figure 1.7. *Tree representation of the politico-strategic context*

In addition to the hierarchy of the components of the landscape, the reconstruction of the essence of the interactions completes the representation of the strategic landscape in the form of a true system.

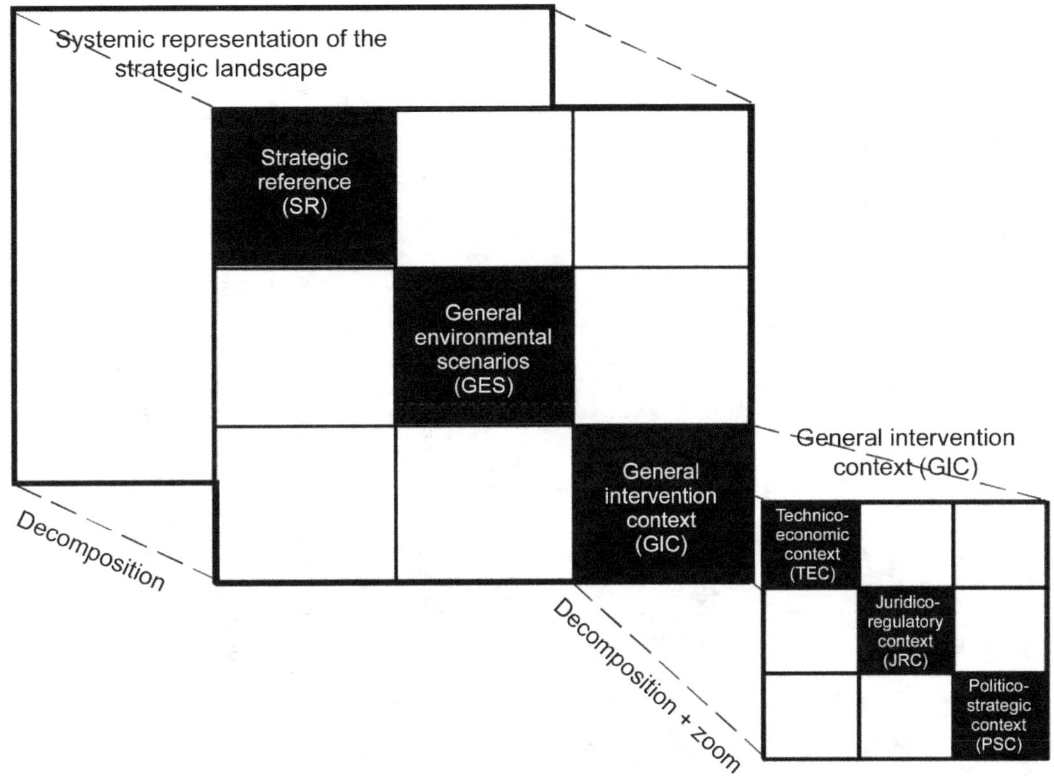

Figure 1.8. *Nested frames and fractal breakdown of the strategic landscape*

1.2.3.2. *Rise of interactions and systemic representation of the landscape*

Reconstructing the interactions of the components at different scales also makes the strategic landscape more intelligible.

The general context of the intervention – and in particular the relationships between actors – lends itself perfectly to such a reconstruction, as it benefits from an aspect that is concrete and immediate.

This method makes it possible to discern the mutual influences of the actors, to detect synergies and antagonisms and even potential alliances and conflicts.

To make the analysis more refined, the bilateral influences or ascending actors from the actors must be expressed in qualitative and then quantitative terms.

The systematic enrichment of the actors' interrelations can lead to the revision of information concerning the protagonists involved.

This sometimes leads to a new understanding of the overall component, located at the higher level.

This highlighting becomes possible thanks to the emergence of a few lines of force governing the actors at the lower level of the encompassing component.

The reconstitution of influences begins with the actors located at the finest levels, and then, in increasing steps, reaches the actors at the intermediate levels, up to the level of the politico-strategic context.

This propagation through stages can be extended to a higher level, that of mutual interactions between the different components of the general context of intervention, of which the politico-strategic context forms part.

This similarity in the representation of the politico-strategic context at different levels is a property of the "structural invariance of scale" or "internal homothety" which, according to its inventor Benoît Mandelbrot, is a characteristic of fractal geometry (see Box 1.5).

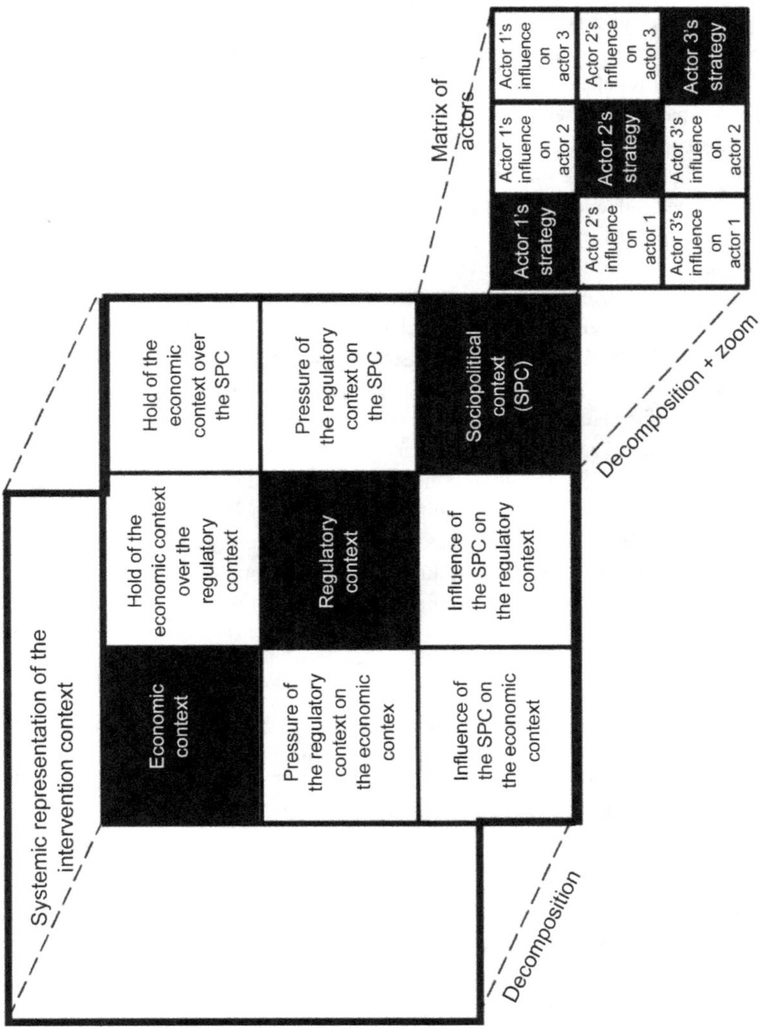

Figure 1.9. *Breakdown of the intervention context into nested frames*

Fractal geometry was discovered by Benoît Mandelbrot, a French–American mathematician (Mandelbrot 1975), in the 1960s, a discovery made unexpectedly while trying to analyze the price of cotton over a period of about 60 years. By examining the curves formed by the prices, he realized that those showing daily fluctuations had roughly the same shape as those illustrating monthly price changes. In his view, it did not matter whether the causes of the price fluctuations were wars, depressions, or the weather, because the degree of recurring irregularity remained the same regardless of the time scale. A fractal pattern is a simple pattern (a shape, transformation, reasoning, etc.) which, if repeated a sufficient number of times, creates an image of high complexity and often of great beauty. Thus, Mandelbrot highlighted the propensity in nature for what can be called the "Russian doll effect"; that is, the production of similar shapes on a very large scale.

Finally, the term "fractal" has become synonymous with a way of conceptualizing, representing, or describing irregular, fragmented, and jagged shapes, from the crystalline outline of snowflakes, to the repetitive shapes of clouds in the sky regardless of the distance from which they are observed, to the blood vessels whose branches make it possible to supply blood cells to the even smallest parts of the body, and up to the lungs, which must contain a very large surface area (larger than that of a tennis court) within a small volume. At first glance, the idea of coherence on increasingly small scales, from a reductionist science perspective, seems to lead to a loss of information. However, the power of the scale invariance that characterizes fractalization manifests itself at a much higher level, given its involvement in the understanding of things at a comprehensive level. Mandelbrot acknowledged that he tried to educate his intuition to accept forms as self-evident that, although widespread, could be considered absurd. For example, in the case of the path of a lightning bolt, it is less the direction than the distribution of its "zigs" and "zags" that is important.

In most of these strange forms, the roughness and entanglements are often the key to unlocking regularities and reaching the essence of the phenomena.

Box 1.5. *The fractal approach: a geometry of nature and thought*

Taking this systemic modeling procedure further, the representation of the politico-strategic context of the intervention can be achieved through a series of frames (square tables) that fit together like Russian dolls. According to the need to deepen the analysis, this allows us to:

– take a finer approach to the representation at different scales of the subsystems that make up this context using a canonic method;

– decipher the essence of this context, by equipping ourselves with a healthy reading grid adapted to its complexity.

This method is then applied to the other two components of the strategic landscape, namely the strategic reference framework and the general future environment.

Each contextual framework encompasses those located at a lower scale, hence the notion of the "self-similarity" of the global analytical framework which constitutes one of its essential characteristics: its fractal geometry.

The reconstructions of the internal interscale systems of each of the components of the strategic landscape – that is the general context of intervention, the future environment, and the strategic reference – form the outline of a dynamic metasystem. This reveals the interactions, interconnections, and structures constituting the strategic landscape as a whole.

The overall view of the wider context of the intervention is carried out with different "depths of field" depending on the subsystems and/or interfaces considered. This differentiation, which remains the responsibility of the analyst, allows us to focus attention on certain particular subsystems located at the core of the problem or concerns being studied. However, it ensures that these subsystems are both:

– subsumed in the overall intervention system, as an integral part of it;

– compared and confronted with other subsystems, which may be less in-depth, because they are considered to be auxiliary.

Therefore, it is through simulating the immersion of the strategic interventions envisaged in their context that their "congruence" is evaluated, which is to say, their ability to integrate into it. A summary retrospective analysis provides an initial validation of the system thus reconstituted.

1.2.3.3. *Elements of strategic vigilance*

The integrated environmental framework makes it possible to lay the foundations of an *ad hoc* monitoring or strategic vigilance system. Indeed, beyond the initial snapshot of the state of the system and its problems, it constitutes an important input for permanent observation and continuous deciphering of the evolution of this environment.

In this perspective, it will help to detect signals – sometimes barely perceptible – reflecting the possible redistribution of power relations between actors, and will make it possible to:

– focus attention in an attempt to reduce, *a posteriori,* the few remaining gray areas around the behaviors of certain actors;

– update, gradually and cumulatively, but in real time, the real data on the environment. The result of this is a better understanding of the major issues facing the company or organization, as well as its emerging challenges, hence its ability to react in phase;

– acquire a capability of anticipation thanks to the pragmatic appropriation of the notion of early warning, and therefore a precious area of intelligence on the context of action, which can facilitate the management of compromises.

1.2.4. *Configuring the intervention as a system connected to the landscape*

Strategic intervention is a coherent set of areas or lines of action that respond to specific imperatives or objectives. This representation of the intervention is compatible with the fact that it integrates a hierarchy of objectives, which can be visualized in the form of a tree whose final objective is located *above,* and whose most operational objectives are located below.

Fundamentally, the strategic intervention should not simply bring together a set of objectives – even if they are finalized – but should identify a system of objectives. These objectives interact and these interactions are an integral part of the system. Thus, in accordance with the principle of the similarity of the nested frames to one another, the project is no longer represented by a simple block of qualitative information, but by a two-entry table (located at the bottom right of Figure 1.10), with diagonal boxes corresponding to the objectives or components of the strategic intervention and non-diagonal boxes to the possible interactions (synergies or contradictions) between them, considered side by side.

1.2.5. *Specifying the interactions of the "landscape" and "intervention" systems*

The interfaces between the actors of the politico-strategic context and the objectives of the strategic intervention are no longer expressed in the form of vectors (columns or rows), but through matrices, the elements of which result from the contrasting (crossing) of each actor of the politico-strategic context with each of the objectives of the strategic intervention.

The assessment of the usefulness and the politico-strategic scope of the intervention can be synthesized and then organized in a matrix (objectives/actors), where each box contains the advantages (approvals) and/or disadvantages generated by an objective-line on an actor-column.

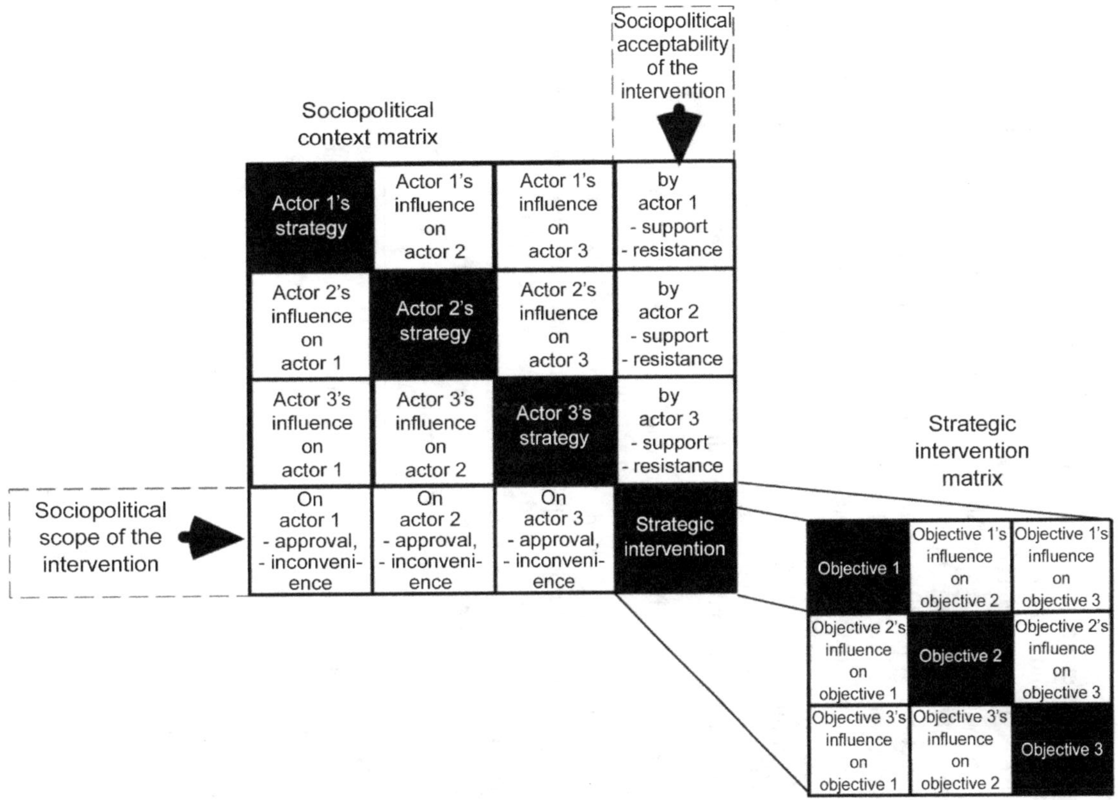

Figure 1.10. *Grid for analyzing the congruence of the intervention within its sociopolitical context*

In the same sense, the assessment of the politico-strategic acceptability of the strategic intervention can be summarized in a matrix (objective × actors), where each box expresses the support and/or resistance of a line-actor with regard to an objective-column.

If most of the methods of analysis of the interplay of actors focus on the matrix of their strategies, and rightly so, they still only provide a partial understanding of the objective × actors' "interface". These analyses rarely focus on the matrix of interactions between objectives. In fact, they only consider a single matrix of objective × actors, the qualitative content of which does not allow us to distinguish the meaning of the interaction. It is difficult to determine whether it is a question of the "usefulness" aspect of the intervention for its context, which comes under the notion of impact, or its "acceptability" through its politico-strategic context, which refers to the notion of environmental control. It would seem that these analyses focus more on the acceptability and reaction of the actors with respect to the objectives, which would make it possible to counterbalance the self-centered tendency of project designers to be interested only in the usefulness of the intervention.

It should also be noted that the systematic development of this matrix can be a substantial step forward in the design of the program and the evaluation of intervention projects. This was the impetus for the methodological initiative advocated in 1997 by the World Bank for the formulation and monitoring/evaluation of programs. This initiative, called the Integrated Development Framework – presented as a true breakthrough in the development approach – was truly noteworthy. In essence, it was limited to the systematization and use of a matrix for comparing the major areas of action envisaged for the main actors involved (see Box 1.6).

According to the President of the World Bank, James D. Wolfenson (1995–2005), the macroeconomic approach to development has shown its shortcomings, thus the urgent need to broaden the framework of analysis to other dimensions, to prevent prescriptions from being ineffective.

To do so, he proposed an integrated framework based on a more global and concerted vision of development. The indicative dimensions of the development strategies to be taken into account within the country framework naturally concern the so-called "structural" aspects (governance, legal/institutional and economic/financial systems, social programs), the human aspects (education and knowledge, health and population), the material aspects (water and sanitation, energy, transport and telecommunications, sustainable development) and the strategic aspects (rural strategy, urban strategy, stimulation of the private sector, etc.). While this extended list of strategic areas to be covered seems generic and traditional, the major interest of the recommended framework lies in the systematic comparison of the activities of the various actors in order to highlight their degree of contribution to the development process.

The main actors concerned can be grouped into governments (main bodies at the central and decentralized levels), multilateral and bilateral agencies (the IMF, United Nations agencies, the European Union, etc.), civil society in all its forms (religious organizations, foundations, trade unions and employee associations, employers' organizations, NGOs, project groups, organizations of indigenous populations, etc.), and the national and foreign private sector.

According to the President of the World Bank, this matrix of comparisons between strategies and actors should be an instrument under the supervision of the government to intensify cooperation, ensure greater transparency, and strengthen partnership. It should serve as a basis for coordinating efforts and for assessing the effectiveness of programs and strategies. The matrix is a synthetic development management tool, opening up cascades of detailed annexes for each box. The integrated development plans would be implemented at a pace specific to each country involved and would have to be kept up to date in real time.

Box 1.6. *The new integrated development framework (World Bank 1999)*

Analyses of the interplay between actors generally only take into account the dominant nature, either favorable or unfavorable, of the actor's attitude towards the objective. However, such a restriction leads to the ambivalent character of the actors as it emerges from the analysis: a given actor may, in certain cases, be both favorable (providing support) and unfavorable (generating resistance) to the same specific objective and, *a fortiori,* to the strategic intervention as a whole. Of course, these analyses make it possible to assess first the structural balance of power between the actors, independently of the objectives, and second the synergies or antagonisms induced by the stance adopted by the actors with respect to the objectives. Nevertheless, these are unable to show:

– the way in which the relations between actors, given that they result from their positions with regard to the different objectives, interfere with the structural power relations of these actors;

– the extent to which these interactions that collide tend to distort or even reconfigure these structured relationships.

The interpretation of the strategic landscape, which initially appears to be rather unstructured, is the first indispensable phase of any strategic analysis. It makes it possible to describe an intriguing and relatively indescribable situation, as these are precisely the most essential characteristics of social situations and human behavior.

2

Imagining and Directing Plausible Futures

The construction of scenarios is a crucial component in the strategic engineering process, in particular the "exploration of prospectives". There is a global consensus on how to build scenarios.

The purpose of this chapter is to provide both a conceptual and practical overview of this important phase of the process by first emphasizing its most important prerequisites, as they involve characterizing the factors of change and identifying major uncertainties (section 2.1). Then, we address the construction of complex scenarios using a two-step morphological analysis procedure. This is followed by the step of *fleshing out* and enriching the scenarios in order to make them more evocative and appealing to decision-makers (section 2.2). A discussion of the issues related to updating and monitoring the scenarios and their use for strategic purposes is also included.

2.1. Exploring the future through the use of scenarios

This section recalls the distinction between foresight and strategy. It reorients the scenarios as constituent elements of prospective exploration and lays out the different components of such scenarios (see Figure 2.1).

The purpose of foresight is to mark the spaces for possible initiatives of the organization concerned, in the form of exploratory scenarios, as they emerge from the evolution of the general environment and its particular context. This strategy, more normative in nature, aims to opt for "long-term positioning", particularly in light of the scenarios. Such scenarios exclude, at this stage of the strategic engineering process, the field of investigation of exploratory foresight. However, for essentially pedagogical reasons, this exclusion is only temporary, insofar as foresight and strategy will then have to be combined to guarantee the adequacy and effectiveness of the action.

2.1.1. *Reorienting scenarios as a core element of foresight*

There are several constitutive aspects of exploratory foresight, as shown in Figure 2.1.

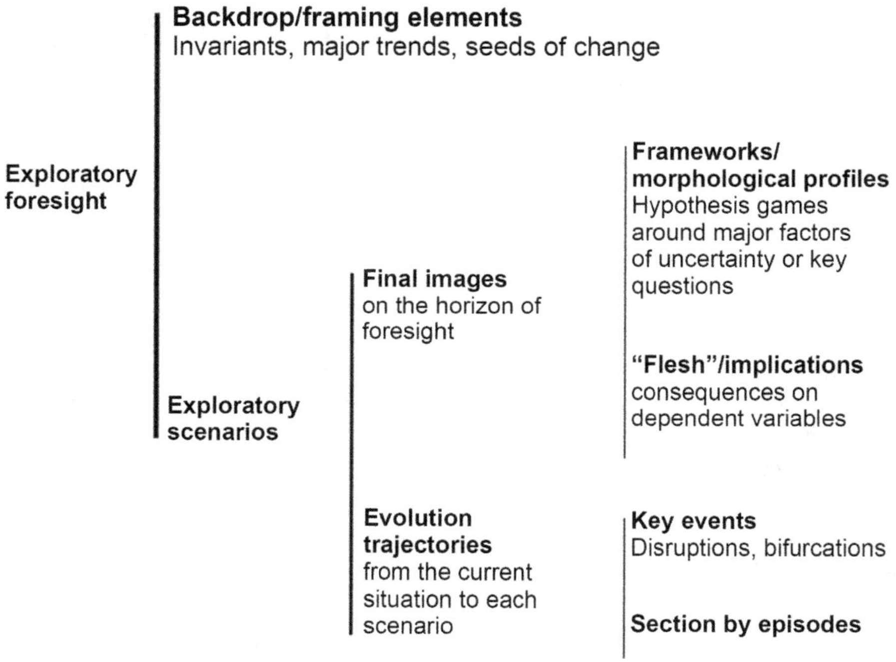

Figure 2.1. *Fields of investigation of exploratory foresight*

The first contains the main generic elements. These are the invariant and strong trends that, in one way or another, form the backdrop of the scenarios. From this common basis, the scenarios will be sketched out, or at least the elements that should govern their development. These elements are identified on the basis of the factors framing the environment, whose evolution can be considered to be predetermined. Such elements, which are most often based on forecasting methods, play an even more important role in exploratory foresight, as this environment evolves without discontinuities. This is why much foresight work was limited during the *Trente Glorieuses* (the "Thirty Glorious Years", which in France refers to the

30 years following the end of the Second World War) to describing unchanging factors and slow-changing tendencies, as well as noting the seeds of change, knowing that these did not have the power to discriminate on alternatives of the future. However, these framing factors do not exclude the consideration of certain changes or ruptures, provided that they are recognized as inevitable in the horizon being explored.

The second aspect concerns the exploratory scenarios which are sketched out as so many diversified forms or alternative reliefs, with the possibility of emerging within the backdrop of the prospective.

In these scenarios, a distinction must be made between “freeze frames”, which describe the possible futures on the time horizon of the prospective study, and the sequence of events, or even the breakages that had to be endured in order to reach them.

As regards the final images of the scenarios, they are based firstly on a “framework”, or morphological profile, whose creation constitutes in substance the essence of the scenario method. This method is interconnected around pivot points, from which the system may, in the future, be brought to switch to alternative configurations.

To be expressive, these scenarios are “fleshed out”; that is, given specifications that make them expressive and more meaningful: more alive. With this same concern in mind, the poet Paul Éluard (1937) said: “It takes few words to express what is essential; it takes so many words to make it more real.” This “flesh” includes the breakdown of the implications of the sets of hypotheses, structuring the scenarios at the level of the dimensions or aspects deemed to be dependent on the evolution of the system.

2.1.2. *Building frames or scenario profiles*

Scenarios are relevant, coherent, and plausible images of possible futures. The approach to building scenarios is essentially based on morphological analysis[1], the different stages of which should be reviewed here. In order to be correctly practiced, it requires a certain number of precautions, of which it is important to make practitioners aware.

1 Morphological analysis was developed by Fritz Zwicky (Zwicky 1962). It gave rise to university research applied to foresight in the 1970s and was rediscovered in France at the end of the 1980s. Since then, it has been widely used in the context of numerous prospective reflection processes.

2.1.2.1. *Identification of major uncertainties and key issues*

The first step of this method consists of "clearing the subject", by reviewing all its facets, then identifying and putting into perspective all the factors and variables that are likely to affect the strategic landscape on the horizon being explored. The second step is to identify, among these variables, those that are demonstrated to provide both structuring and discernment in describing the future.

In order to select these variables, they are filtered through a cascading sorting process based on three key elements or criteria that refer to, respectively, the notions of necessity, willingness, and uncertainty:

– the first criterion is the *importance* of variables, that is, their structuring and unavoidable, and therefore necessary, ability to imagine and describe the future;

– the second criterion is *control*, and concerns the presumed capacity to reasonably exercise one's will in order to pivot or control the progression of variables in the desired direction;

– the third criterion is *uncertainty*, which takes into account the difficulty one feels in considering the progression of these variables as predetermined. It therefore reflects the plurality and unpredictability of the trends that these variables conceal.

Figure 2.2 presents the path taken in the processing of variables and factors in a strategic foresight approach.

The field of prospective exploration involves variables and factors that are both important and uncontrollable. These are factors affecting the general environment and the future context. These factors are therefore likely to impose themselves on us, and we will therefore be influenced by them, with varying degrees of success. In this field, the most natural process of anticipating the future requires us to be able to rely on benchmarks. Some of these allow us to rely on stable or predetermined elements, such as unchanging factors and strong tendencies, which constitute the basis of the future, the necessary basis for the construction of the scenarios. These benchmarks also include the pivot points around which the range of possibilities unfolds and/or on which the future could hinge.

The construction of the exploratory scenarios will therefore be built around these pivot points, described as "major factors of uncertainty" because of their consequential, uncontrollable, and discriminating nature. As most global crises insistently remind us, and in the words of Al Gore (2018): "The drama does not come from uncertainty as such, but from what is held to be certain and which is not." This is why it is necessary to ensure the rigorous application of the law, of the third sorting criterion, in order to distinguish between major trends that are relatively predictable and regular and major uncertainties, and to avoid exposing oneself to crucial strategic risks.

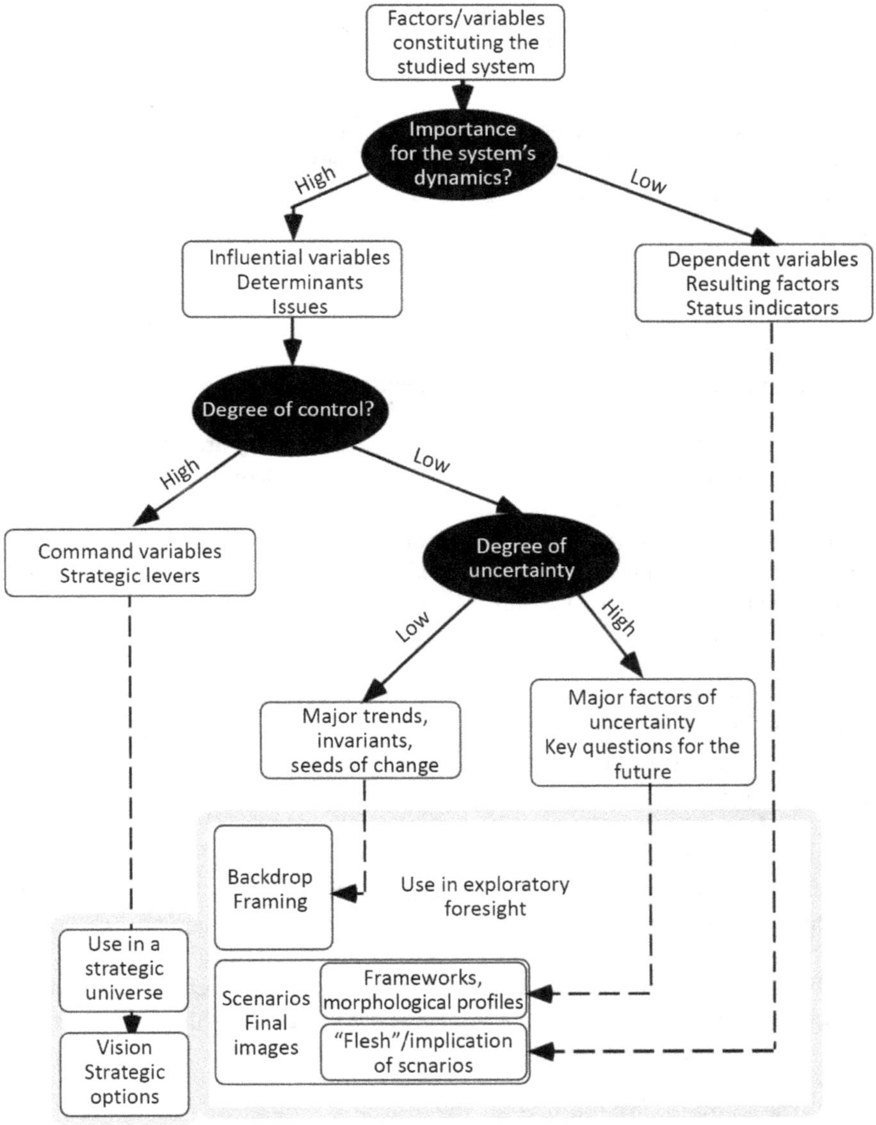

Figure 2.2. *Classification and use of factors/variables in strategic foresight*

2.1.2.2. *Importance of sorting variables beyond the exploratory field*

Referring again to the cascading organization structure presented in Figure 2.2, important factors and variables that are controllable – that is, that one is able to control, for example, by influencing their course of change – fall within the scope of strategic action.

Finally, the important factors and variables which are not within the scope of prospective exploration or strategic action are part of an intermediate field of primary interest: the vision or strategic plan discussed in Chapter 1. This vision is more precisely connected around the crucial issues, the major trade-offs, and the alternatives decided upon in terms of the major challenges to be met.

Exploratory scenarios can serve not only as a source of inspiration for visionary and strategic action, but more importantly, as criteria for a contextualized assessment of the long-term validity of different options.

Although the three fields of prospective exploration, proactive vision, and strategic action are interrelated, both in terms of methods (morphological analysis) and substance, they are nonetheless different in terms of the factors and variables they involve. For example, it would be illusory to claim to equate a strategy with the exclusive choice and appearance of a desirable scenario. This is because, as shown in our flowchart (Figure 2.2), scenarios and strategies do not apply to the same categories of factor variables.

The variables involved in foresight are, by their very nature, beyond the control of those who develop them. Indeed, they cannot be subject to the deliberate will of decision-makers, except in a relatively marginal way, through the strength of their strategic vision.

These considerations lead to realism or reasoned pragmatism, the first imperative of which is to ensure the quality of the categorization made between the variables and their distribution between the fields. Emperor Marcus Aurelius (121–180 A.D.), who would have already perceived the acuteness of this critical issue in his time, declared: "God, give me the serenity to accept what I cannot change [the field of prospective exploration], the courage to change the things I can [the field of strategic action] and the wisdom to know the difference [the sorting keys]."

2.1.2.3. *Development of the scenario framework*

Scenarios are constructed using the morphological analysis approach, as shown in Figure 2.3:

– the first step of the analysis involves the identification and transcription of the major uncertainties, in the form of key questions about the future;

– the second stage of the analysis consists of imagining, alternative hypotheses for each area of uncertainty that are both plausible and mutually exclusive, in the form of contrasting answers to the key questions raised. These are represented by white bricks in Figure 2.3;

– the third step of the analysis consists of combining a few contrasting profiles of hypotheses in a coherent manner and for all uncertainties, which represent the exploratory scenarios.

Approach	Outline	
Major uncertainties or **key questions for the future**	Major uncertainty 1. *Key question 1.* Major uncertainty 2. *Key question 2.* Major uncertainty 3. *Key question 3.*	
Hypothetical situations or **modalities of responses**	Key question 1. Key question 2. Key question 3.	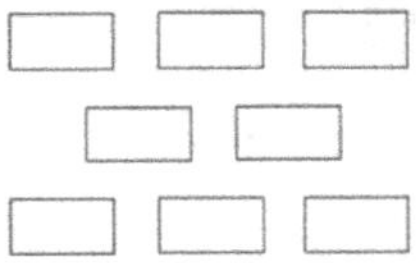
Combinations of hypotheses and profiles of **exploratory scenarios**	Key question 1. Key question 2. Key question 3. Scenarios	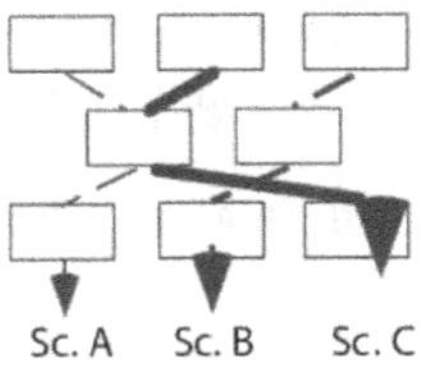

Figure 2.3. *Major uncertainties and scenarios: the stages of construction in Adobe Illustrator format*

2.1.3. *Sketching morphological profiles*

A morphological profile (or combination of alternatives) is materialized by the drawing of a line of segments connecting the contrasting response alternatives. It is advisable to carry out this plotting by means of a cascading progression from the dimensions of uncertainty corresponding to the upper lines providing the most structuring and influential response alternatives to those deemed to provide the least structuring alternatives.

After considering the question of the coherence of the scenarios preselected in this way, we select one scenario considered most likely, while also examining two or three that are less significant but with more contrast, in an effort to open up the field of possible futures. These scenarios are constructed in an exploratory manner. Nevertheless, they make it possible to mark out areas of freedom in the medium term for the main actors involved in the evolution of the context of the company or institution concerned.

The purpose of this stage of prospective reflection is therefore to provide a prospective framework for strategic actions, including a backdrop that includes areas of inertia and major trends, setting the stage for the various exploratory scenarios selected that will be sketched out.

2.1.3.1. *Identification and grouping of important factors*

At this stage, it is appropriate to insist on the usefulness of a structural analysis as a tool for forming the integrated base of information – in this case, all the factors and variables – necessary for the prospective exploration and the construction of scenarios.

The method of structural analysis consists of a cognitive scanning of the system to be studied. The purpose of this method is to identify the focal points of the problem, which constitute either major uncertainties or future challenges, around which different sets of actors can be connected. The emphasis is not on a structural analysis as such – and one considered to be known elsewhere – but on its interpretation, with the goal of developing exploratory scenarios.

2.1.3.2. *Stratification by categories of uncertainty factors*

The cascading of the scenario profiles is all the more complex given the large number of key dimensions. To overcome this difficulty, the list of key dimensions should be split into strata:

– The first layer, which relates to the backbone of the scenarios, will be located in the upper part of the canvas. This will bring together the subset concerning the environment, which will include the dimensions of uncertainty considered to be the most global and structuring. It is around these dimensions that the framework conditions, which constitute the "morphological space" of the global environment scenarios, will be connected.

– The second layer, what we call the "cartilage" of the scenarios, will include the dimensions considered less important in providing structure. These will be the uncertainties relating to the more specific (or close) context of the problem under study. Thus, the fact that such-and-such a configuration is created will be subject to the upstream knowledge of the global environment scenarios. Within the framework of a fixed general environment scenario, it is advisable to extend the route so that the second layer is subject to the first. It will then be necessary to identify the most appropriate combinations, dimension by dimension, while ensuring the internal coherence of their arrangement. To return to our metaphor, at this stage, an attempt is made to increase the surface area of the prefabricated wall sections "at the margin" in certain places by adding bricks. The arrangement of such bricks is, by construction itself, linked to the existence of the original wall.

– The third and last layer, including the auxiliary dimensions, corresponds to what we call the "flesh" of the scenarios, or to the plaster on a wall.

Once the principle of hierarchical ranking of dimensions is accepted, the question arises as to where the boundaries of each of the dimensions of uncertainty covered by these three strata lie. This question is crucial because the result of scenarios, for the given *ad hoc* reflection group, will be affected by the trade-offs that will be made, both upstream of the process and at different points in its progress.

2.1.4. *Adopting a two-step procedure*[2]

To reduce the complexity of the foresight process, it is possible to systematize the determination and selection of the scenario profiles by adopting a two-step procedure, summarized in Box 2.1.

> To facilitate the construction of the scenarios, a two-level procedure is applied.
>
> The *first level* concerns the creation of *scenarios specific to each of the adopted themes*. The tables summarize the sets of hypotheses (concerning the responses) underlying the different thematic scenarios.
>
> The *second level* concerns the *consolidation* of these in the form of global scenarios.
>
> At this stage, it is no longer a matter of combining the elementary answers, but of combining the thematic scenarios, which are then considered as composite hypotheses. This

2 As far as we know, the first approach to have used the principle of the "two-step morphological analysis" since the early 1990s was *La Catalogne à l'horizon 2010*, led by Hugues de Jouvenel. Among the companies and organizations that have used the multilevel morphological analysis approach with our methodological support since 1993, we may list: the Sirius/Idate laboratory for prospective telecommunications, La Poste (the Postal Service of France), Crin, Datar, France Télécom, Axa, TDF, Usinor Sacilor, GDF, the SNCF (the national railway company of France), the French Ministry of Defense, and the Department of Economic and Social Affairs of the United Nations.

aggregation procedure, in substance, consists of repeating the same method at a more global level:

– *1st floor*: identification of essential dimensions (major uncertainties or key issues) and determination of thematic areas.

– *2nd floor*: construction of the scenario frameworks according to theme:

- *creation of frames for thematic scenarios*: hypothetical situations and the ways of answering each key question,

- *determination of morphological profiles of thematic scenarios*: contrasting sound combinations, one for each dimension,

- *naming and structuring of the thematic scenarios*: vertical alignment of their respective hypotheses.

– *Level 2*: aggregation of thematic scenario frameworks:

- *development of the global scenario framework*: thematic scenarios as composite hypotheses for each theme or macro-dimension,

- *determination of morphological profiles of global scenarios*: contrasting combinations of thematic scenarios,

- *naming and structuring of global scenarios*: vertical alignment of their thematic scenarios or composite hypotheses,

- *presentation of the detailed structure of the general environment scenarios*: highlighting their founding assumptions.

Box 2.1. *Procedure for constructing two-step scenarios*

2.1.4.1. *Sequence of steps*

The procedure consists of classifying the major dimensions of uncertainty concerned into domains or themes, and then considering each of these themes as an integral component of the layer under consideration.

The schematic presentation of the first level of this procedure is shown in Figure 2.4A. It consists of exploring and selecting a few profiles of scenarios for each area or theme, for which reason the terms "fragmented" or "thematic" are used for the profiles, as they are limited to a particular theme.

2.1.4.2. *An example concerning technological research in Europe*

To illustrate this procedure in an educational way, we will use the presentation of certain results of an exercise that we co-directed, at the methodological level, within the framework of the Club Crin Prospective scientifique et technologique[3]. In this example, Figures 2.5A, 2.5B, and 2.5C illustrate the procedure for the scenarios of the "economy" theme alone. Figures 2.6A, 2.6B, 2.7A, and 2.7B summarize the results obtained for all the five themes studied.

The second level of the procedure is shown schematically in Figure 2.4B. It consists of repeating the procedure applied at the first level, but considering the thematic components as macrodimensions of major uncertainties or generic key issues. The *composite* hypotheses associated with this component represent the fragmentary scenarios derived from the previous level for the theme in question.

Thus, a global scenario including all the themes studied is a coherent combination of composite hypotheses, with only one important hypothesis being retained per theme. Figures 2.8A, 2.8B and 2.8C show the development of this second level of procedure, concerning the creation of the general environment scenarios of the previous example. The procedure can be repeated as many times as the level of aggregation of the scenario summaries justifies it. Thus, pushing this procedure over a high number of levels, at least theoretically, will lead to an identical reproduction process of the same method. In accordance with the concept of fractal objects discussed in Chapter 1, this procedure corresponds to a *reverse fractal-type* recomposition of the scenario framework construction.

The construction of the global scenarios could also be carried out through a procedure of direct combination of the configurations resulting from the different framing factors or major uncertainties.

This last approach offers, *a priori,* greater flexibility and more accuracy in taking into account the factors of framing.

In effect, developing global scenarios by assembling "prefabricated" sub-scenarios by field or theme leads to a less satisfactory end result than in the case of direct construction.

On the other hand, the large number of factors taken into account in this last type of assembly makes creating combinations more complex, since this involves identifying and at the same time selecting a few scenarios within a myriad of possibilities. The two-level aggregation method, on the other hand, allows for a cascading – and therefore gradual – selection and appropriation of scenarios.

3 The Crin club is a component of the Ecrin association, which brings together major companies and the most important research organizations in France. This exercise, carried out with our methodological support, resulted in a publication entitled "Entreprises 2010: construire des scénarios d'environnement global", *futuribles.com,* collection TRP, no. 5, 1997.

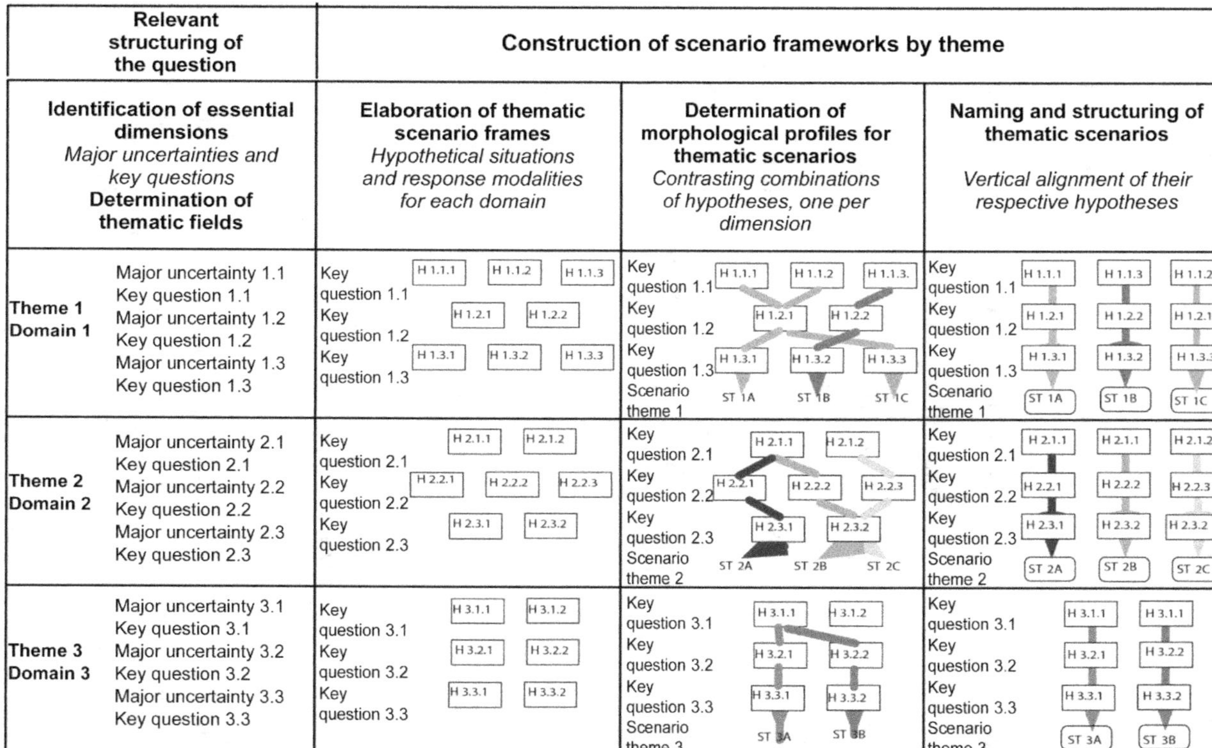

Figure 2.4A. *Development of general environment frameworks or scenario profiles*

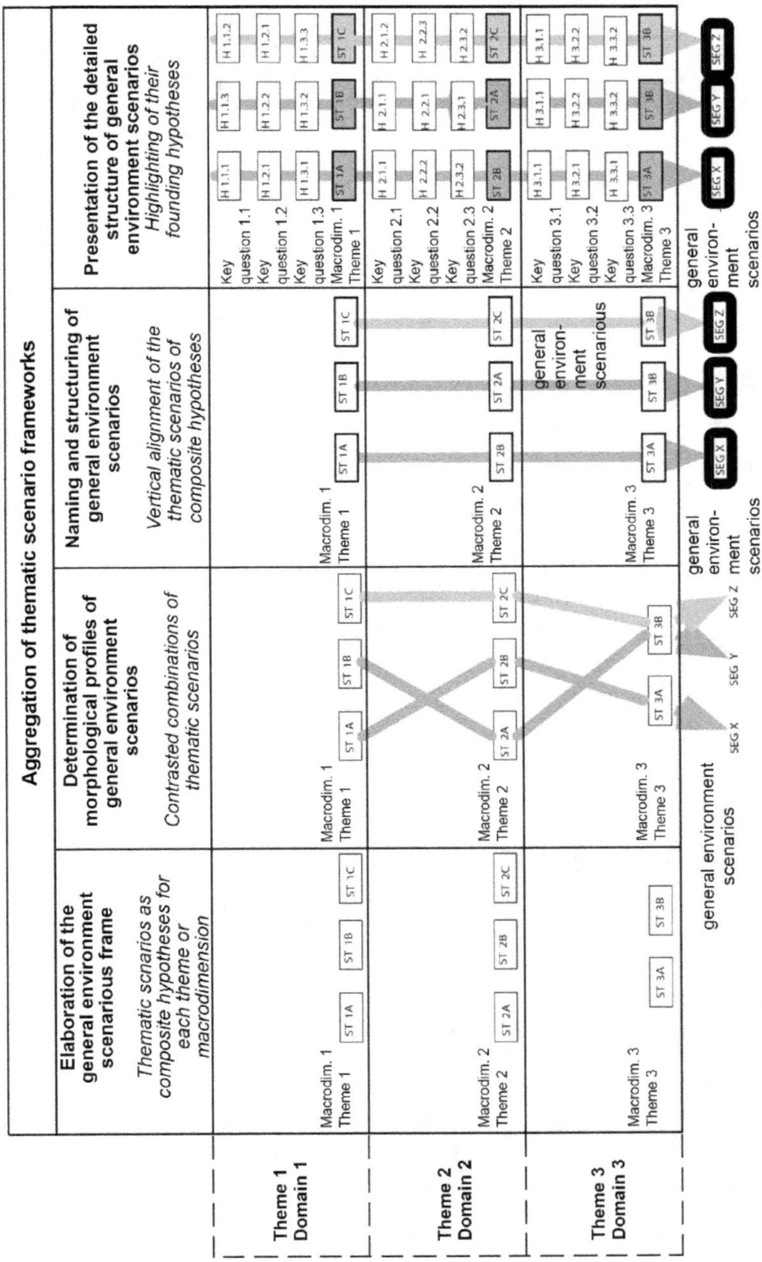

Figure 2.4B. Development of general environment frameworks or scenario profiles (continued)

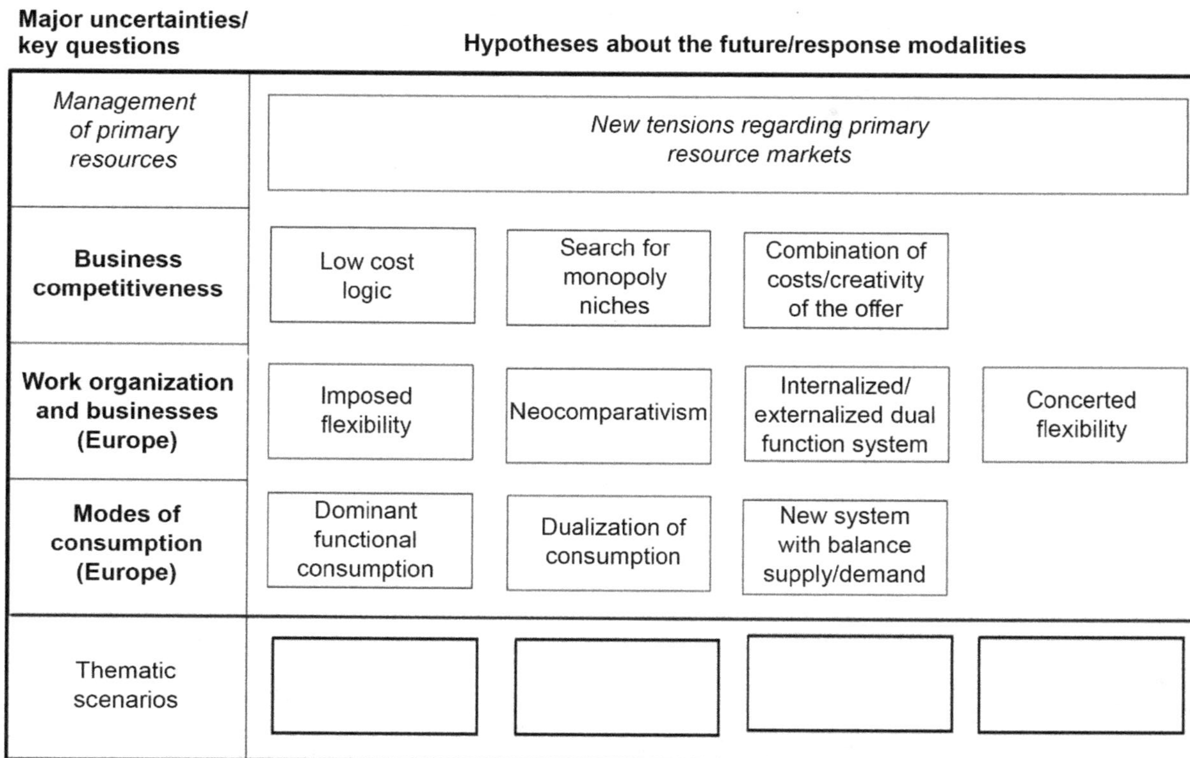

Figure 2.5A. *Economic theme scenario frames (source: Enterprise 2010, Club Crin, TRP No. 5)*

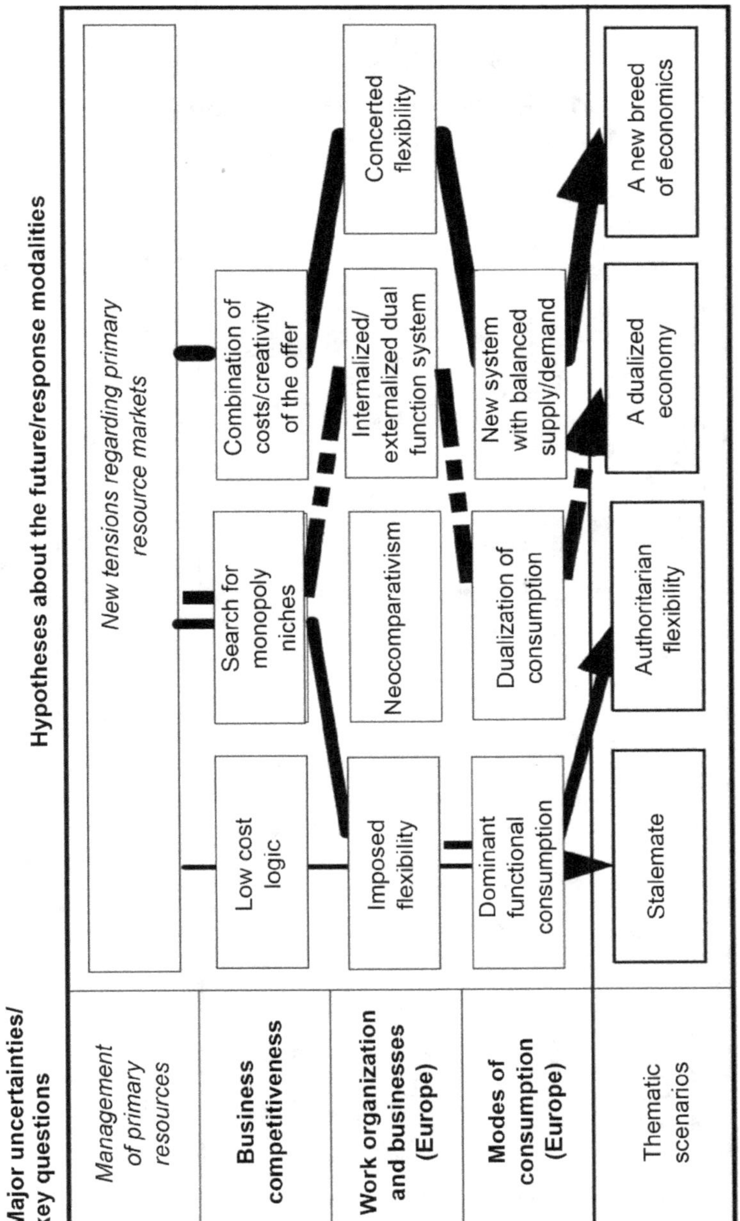

Figure 2.5B. *Economic theme scenario profiles (continued) (source: Enterprise 2010, Club Crin, TRP No. 5)*

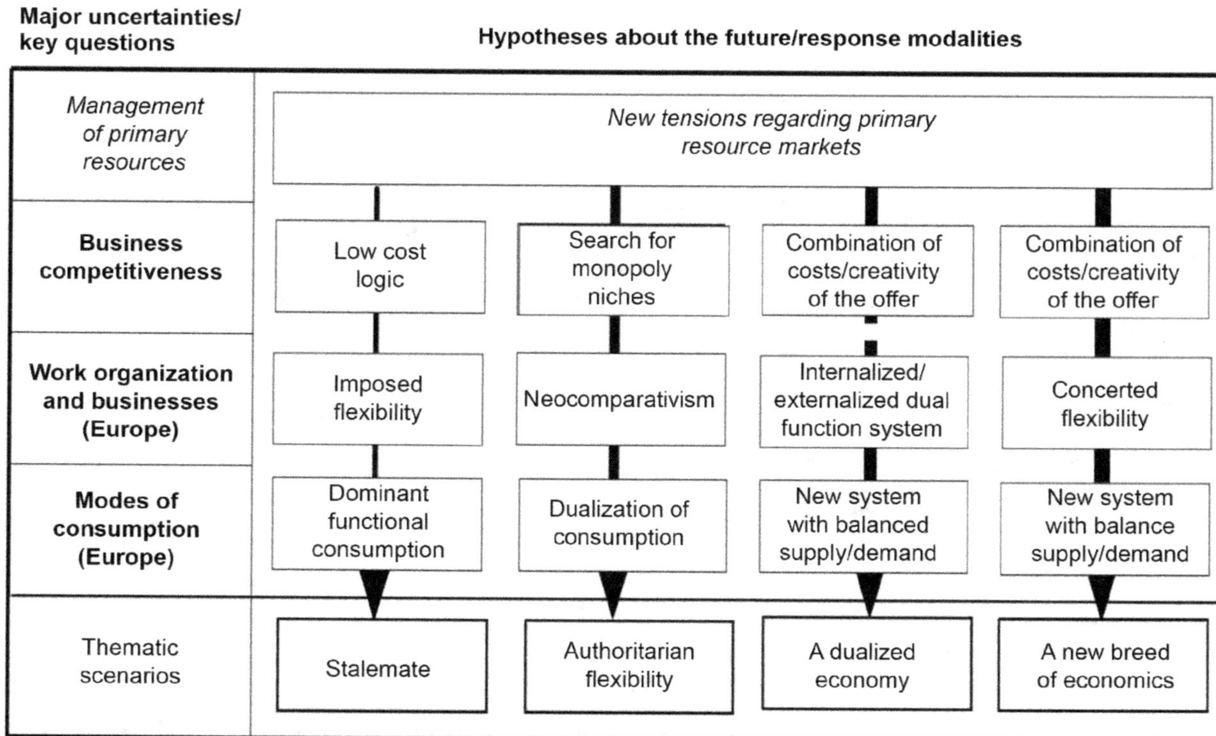

Figure 2.5C. *Summary of economic theme scenarios (continued) (source: Enterprise 2010, Club Crin, TRP No. 5)*

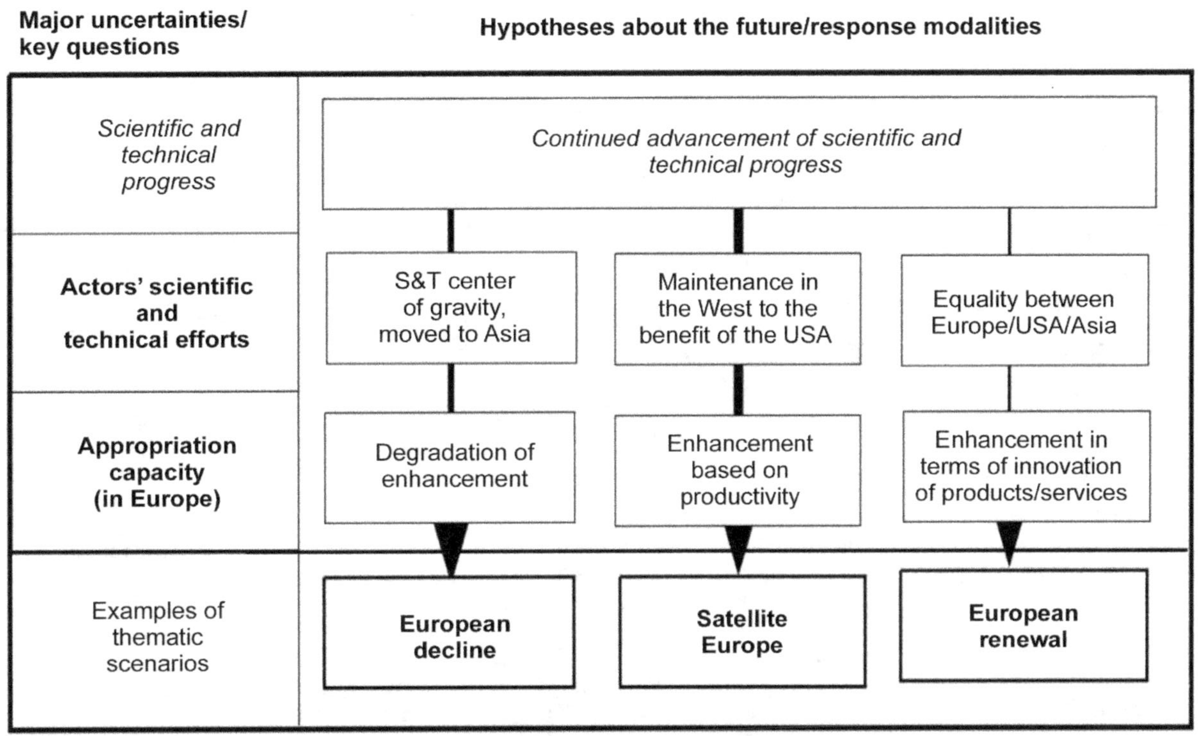

Figure 2.6A. *Summary of science and technology theme scenarios (source: Enterprise 2010, Club Crin, TRP No. 5)*

Major uncertainties/ key questions	Hypotheses about the future/response modalities			
Sociability	Community withdrawal	Community withdrawal	Cohabitation of communities/ networks	Reticular opening
Family structure	Distorted family	Extended blended family	Current maintained mix	Extended blended family
Social cohesion	Aggravated social division	Aggravated social division	Relative current cohesion maintained	Significant reduction of the division
Psycho-sociological tensions	Increased intolerance of anxieties	Increased intolerance of anxieties	Actual level maintained	Managed anxiety
Examples of thematic scenarios	**The social crash**	**The ghetto cocoon**	**Cohabitation**	**Healthy networks**

Figure 2.6B. *Summary of scenarios for the sociology theme (continued) (source: Enterprise 2010, Club Crin, TRP No. 5)*

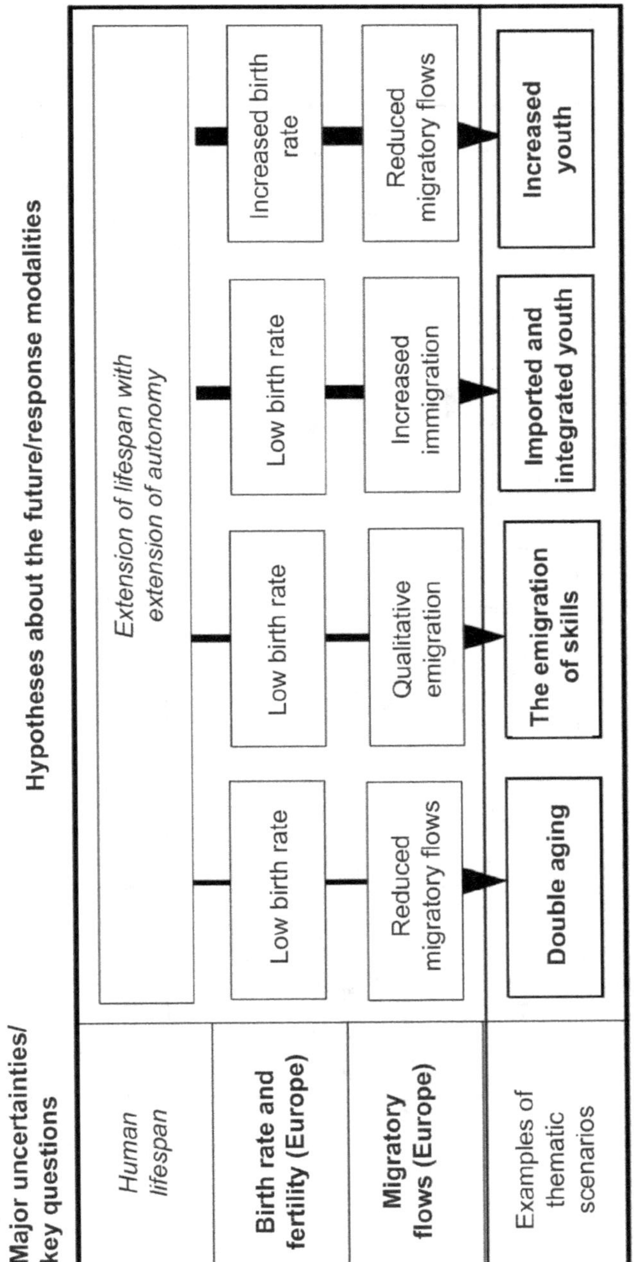

Figure 2.7A. *Summary of demographic theme scenarios (source: Enterprise 2010, Club Crin, TRP No. 5)*

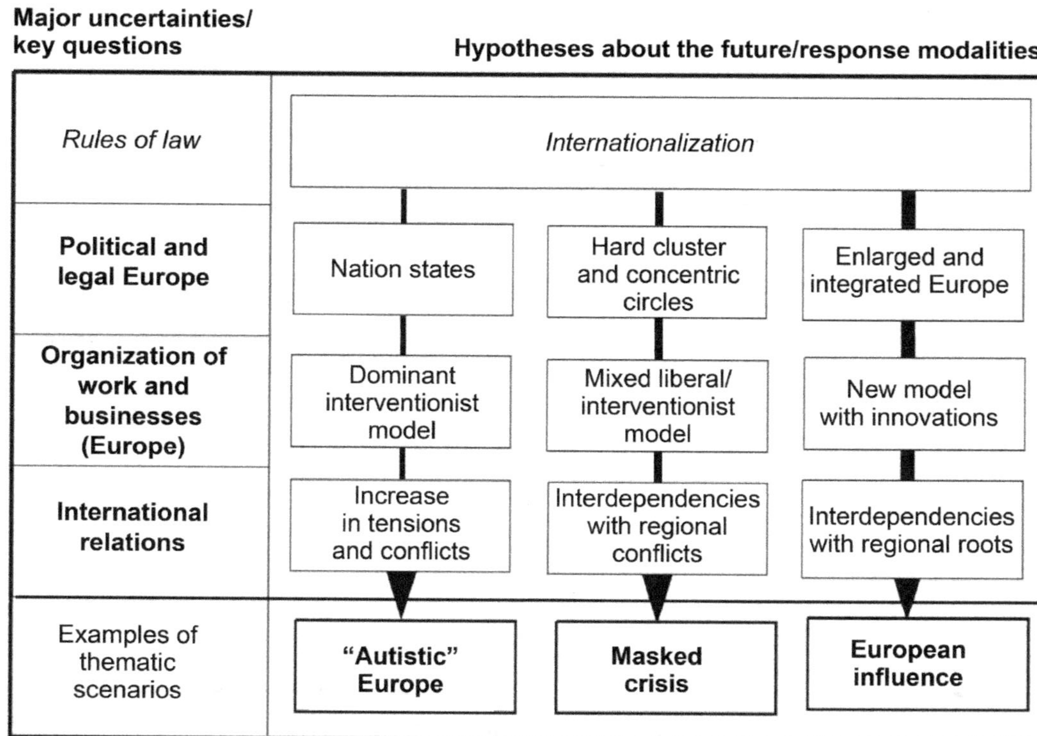

Figure 2.7B. *Summary of scenarios for the legal-institutional and geopolitics themes (continued) (source: Enterprise 2010, Club Crin, TRP No. 5)*

Themes/ macrodimensions	Thematic scenarios/composite hypotheses			
Science and technology	European decline	Satellite Europe	European renewal	
Economy	Stalemate	Authoritarian flexibility	Dualized economy	A new breed of economics
Sociology	The social crash	The ghetto-cocoon	Cohabitation	Healthy networks
Demography	Double aging	Emigration of skills	Imported and integrated youth	Increased youth
Legal-institutional and geopolitical	"Autistic" Europe	Masked crisis	European influence	
General environment scenarios				

Figure 2.8A. *General environment scenario plot (note that the second line of this figure is taken directly from the last line of Figure 2.5C, which summarizes the economy sector) (source: Enterprise 2010, Club Crin, TRP No. 5)*

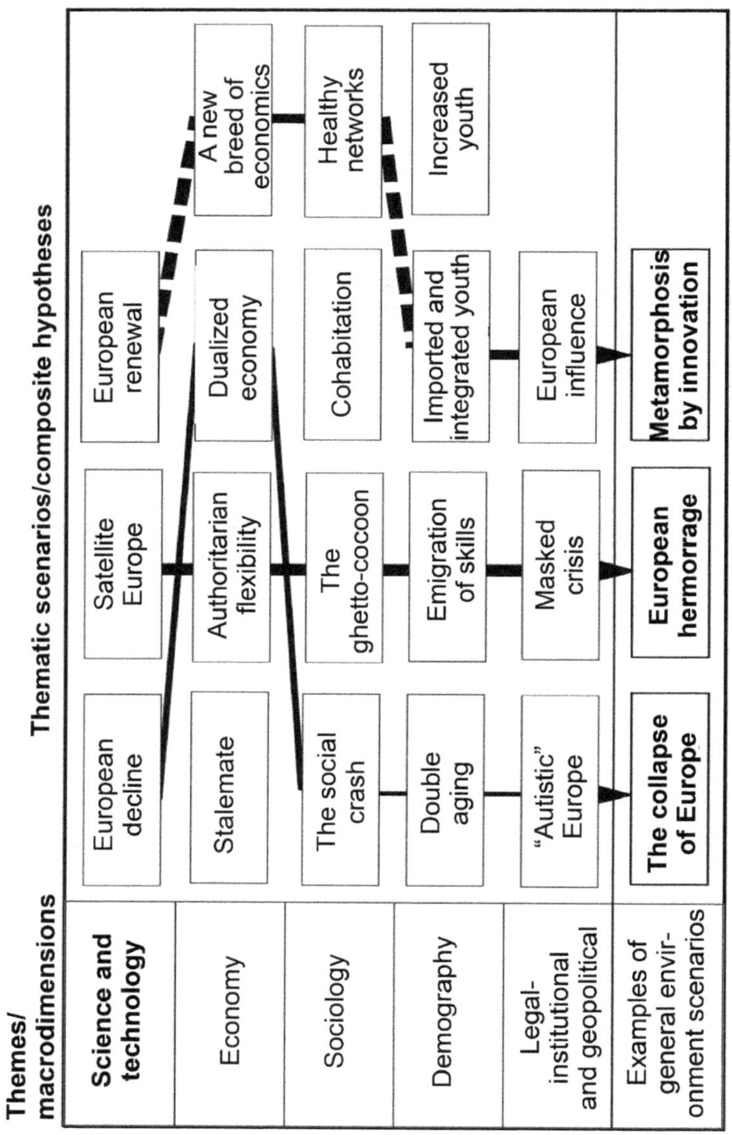

Figure 2.8B. *Profile of the general environment scenarios (continued) (source: Enterprise 2010, Club Crin, TRP No. 5)*

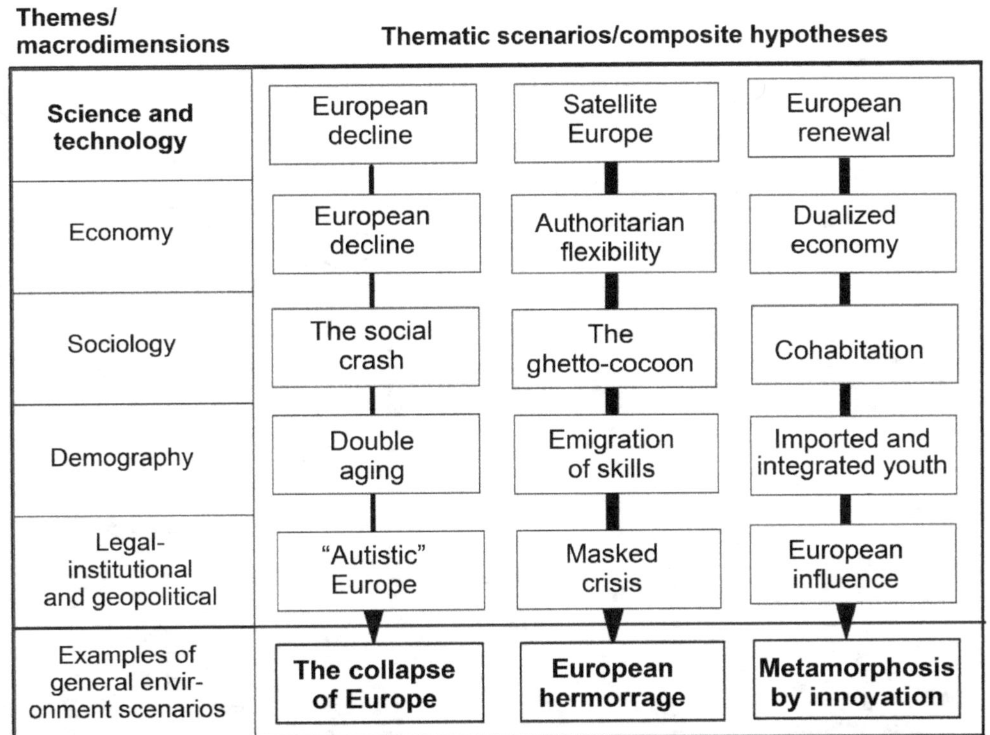

Figure 2.8C. *Summary of the general environment scenarios (continued) (source: Enterprise 2010, Club Crin, TRP No. 5)*

The two-tier process of constructing scenarios is facilitated by the use of the Morpholudis tool, summarized in Box 2.2.

> Morpholudis is an instrument for a morphological analysis using a "playful cognitive" approach. Morpholudis, designed by Jacques Arcade, is intended to help decision-makers and strategists to mix key ideas or combine new concepts to make successful "cocktails" and generate a palette of contrasting alternatives.
>
> More specifically, Morpholudis allows for the creation of either exploratory or prospective scenarios, or strategic options. The method, using a visual input such as a computerized "Rubick's Cube", creates relevant, coherent, and typified combinations of hypotheses relating to the future (in the case of exploratory scenarios) or modes of intervention (in the case of strategic options).
>
> The main results of Morpholudis also focus on a proactive vision and the areas of overachievement or breakthrough, key skills, and core businesses, product and service concepts, R&D, and innovation policy areas.

Box 2.2. *Morpholudis*

Figures 2.8A, 2.8B, and 2.8C show the overall scenarios, with all themes combined, resulting from the application of this method.

2.2. Reinforcing and fleshing out the scenarios

It is possible and desirable to perform a hierarchical breakdown of the exploratory scenarios in the form of a tree structure.

This hierarchical transposition consists in "segmenting" the future according to the scenarios and their founding assumptions. These are presented in Chapter 4. The description of the scenarios in the form of a tree structure is intended to facilitate the consolidation of their framework as presented in section 2.2.1. This organization of the scenarios into hypothesis trees promotes their use as criteria for assessing the long-term validity of the planned actions.

2.2.1. *Enriching and completing global scenarios*

Within a given scenario, the first task is to refine the characterization of the configurations.

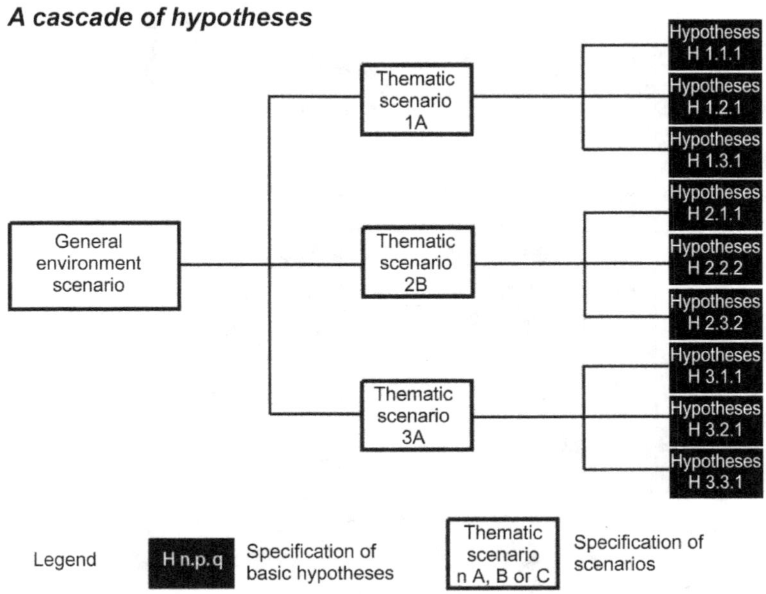

Figure 2.9. *Tree structure representation of a general environment scenario*

This involves the naming of a combination shared between two scenarios – that is, located at the intersection of two plots – according to the profile to which it belongs.

The decomposition of all the elementary (non-composite) configurations of a cross-theme scenario can lead to adjustments in order to ensure the consistency of its profile.

2.2.1.1. *Consolidation of the bones ("the cartilage")*

It is also necessary to consolidate the frames significantly. The approach consists of clarifying the synchronic cohesion of their hypotheses. The goal is to move from a simple system of the combination of hypotheses about the future to a real system of hypotheses. Indeed, this system will be composed of the elements/hypotheses and their relationships (of catalysis, conditioning, influence, etc.).

In instrumental and pragmatic terms, any framework of scenarios should ultimately lead to the completion of a double-entry table of systemic representations, containing as many rows and columns as there are key dimensions, as shown in Figure 2.10.

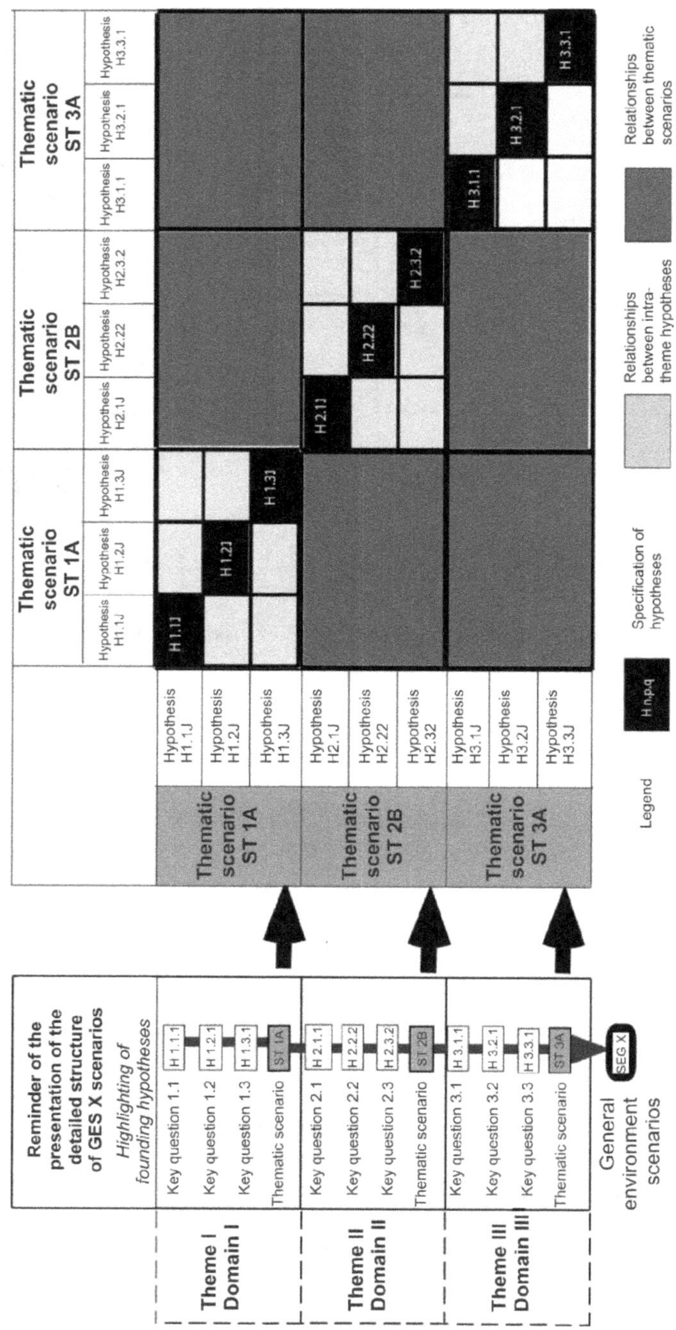

Figure 2.10. *Systemic representation of a general environment scenario*

The assumptions associated with each key dimension, respectively, are shown in the diagonal boxes. The non-diagonal boxes contain items to qualify the conjunction between the crossed hypotheses.

Such a systemic representation of the scenarios allows for their key elements to emerge from their respective hypotheses.

This provides the possibility of identifying the fundamental constituents for each scenario, that is to say, the core of the hypotheses and the interrelationships that characterize it. The fundamental constituents represent a powerful distinguishing element in relation to the other scenarios. Referring to these core hypotheses and relationships will determine the best title, both synthetic and emblematic, to be attributed to the scenario. This often leads to the observation that the initial hierarchy – valid regardless of the scenarios and of a generic nature – of the key dimensions to be taken into account for the identification of the different scenarios finally gives way to specific hierarchies, more in line with "genetic codes".

2.2.1.2. *Covering the frameworks ("fleshing out")*

In order to flesh out the framework of the scenarios, it is important to identify a few additional dimensions that can help to clarify and characterize them. Like the major uncertainties, these additional dimensions of uncertainty are based on the variables or factors examined in the structural analysis. However, they will be more dependent variables. They are considered to be the result of the evolution of the system, thus they are relatively sensitive to the composition of the framework.

For each environmental and/or contextual scenario, it will be necessary to create a breakdown of its implications – by justifying and then describing them – according to each of the selected dimensions. Contrary to the construction of the scenario frames, this exercise for qualifying the impacts does not require the prior constitution of contrasting alternatives because it is a custom-made finishing work. Therefore, this work can be associated with the display of the "plaster" on the walls and the covering of the frameworks with "flesh". This "flesh", added to the frameworks of scenarios, is essential to describe them and to make them credible and "evocative".

2.2.2. *Periodically updating the scenarios*

2.2.2.1. *Monitoring their advent*

In addition to updating the content of the scenarios and their probability of occurrence, monitoring their occurrence involves observing the trajectories that allow them to evolve from the current situation for each of them.

This function involves the rigorous monitoring of relevant and updated indicators. It consists of scrutinizing critical crossroads along the trajectories from which the scenarios emerge, by highlighting key events or verifying their founding hypotheses. It is still necessary to put in place an adequate system and to be able to interpret these indicators with satisfactory reliability in terms of scenarios. Under these conditions, such indicators may be considered as warning signs, or even early alerts, of the fact that we are heading or branching off towards one scenario rather than another. It is more a question of focusing attention on the process of how the scenarios appear, rather than on the *a priori* distribution of the probability that they will occur, even if it is updated. The value of this type of monitoring lies in the possibility, when the situation observed so suggests, of re-evaluating the strategies in light of new information about the scenarios, and revising them if necessary.

2.2.2.2. *Circumstantial review*

The future does not necessarily take the form of one of the scenarios imagined, hence the need to rework or even reinvent them over time. The challenge is to stick as closely as possible to the eventual occurrences in the future in order to be able to deal with the surprises that may be reserved for the company or institution. While the revision process is based on both the method and the results of the scenario-building phase, it is nonetheless delicate. It is therefore necessary to be able to take advantage of the available material while integrating the new data of the future environment (appearance of discontinuities and ruptures, emergence of new dimensions or key questions, exploration of singular response mechanisms). Box 2.3 provides an indication of the main stages of the recommended approach.

1) Explore and identify possible additional questions or dimensions.

2) Classify them according to the degree to which they provide structure (skeleton or "flesh").

3) Identify those that provide the most structure and fall within the framework of the scenarios:

– *for those that are different from the existing key dimensions*, identify their configurations and add them to the frame;

– *for those that are close to existing key dimensions*, reword the question, then revise the configurations, refining them, deleting some and/or adding new ones.

4) Modify the scenario framework accordingly.

5) Re-draw the frames of the scenarios by linking their respective configurations on the modified screen.

6) Concerning the less structuring aspects (questions or dimensions not retained in step 2), describe their configurations on the basis of the framework of the scenarios.

Box 2.3. *Key steps in reviewing scenarios*

2.2.3. *Using scenarios for strategic purposes*

The construction of exploratory scenarios makes it possible to highlight the key issues and major uncertainties that may affect the general environment and/or the specific context.

2.2.3.1. *Application as a source of inspiration for structuring actions*

The scenarios themselves serve as evocative and plausible images of the future context, expressed as sets of hypotheses. When they are appropriated by each actor at different levels of the organization, this has the potential to broaden the range of possibilities. Indeed, this can change the way each actor looks at his or her own long-term space to maneuver, especially in different fields such as marketing, technology, distribution, etc.

Due to its evocative nature, the description of each scenario is a source of inspiration for strategic action. It eases the process of identifying appropriate recommendations. By repeating this exercise for each of the scenarios, stakeholders, and decision-makers can thus formulate a multiplicity of strategic options to cover – at least partially – the many different possible futures.

2.2.3.2. *Use as a criterion for assessing the future validity of actions*

The scenarios have the advantage of challenging decision-makers by encouraging them to examine the reference situations that may be imposed on them. However, decision-makers often have an opinion on the favorability of the scenarios presented to them.

An initial logic applied to the choices made by the company or organization concerned is to select a scenario deemed desirable, while striving to bring it about, including through actions of lobbying. Provided that the company is in a monopoly situation or has "overwhelming" power in its environment, this attitude is unlikely to be realized.

Without underestimating the virtues of voluntarism, it is in the company's interest to rely on the main lines of force of outside elements (economic, regulatory, technological, or social forces, etc.) in order to cause it to progress in the desired direction, or to adopt a long-term position enabling it to make the most of the trends observed.

A second decision-making logic involves the solution adopted as part of a planning process. For the company, this consists of relying on a reference scenario, from among the most probable, according to which it will break down its major orientations. However, the decision to "put all of its strategy eggs in only one basket/scenario" is riskier, since it is impossible to reasonably predict the scenario that will come to pass, despite the monitoring of appropriate indicators.

In any case, the company or organization involved cannot avoid considering the possible surprises or disappointments that may arise in the scenario that will come to pass. It will then have to question the validity of its strategy, in the event of a scenario that breaks with the one it was expecting. To do so, it must compare each strategic option to the different scenarios, trying to assess its long-term relevance in light of these scenarios.

Thus, a third, rather circumspect logic would involve the company or organization doing the following: either distributing its strategic options according to different scenarios by adhering to a policy of spreading and distributing risks, or focusing on a common core of relevant actions, whatever the scenarios. This can be completed, as time progresses, through the close monitoring of how the future unfolds.

3

Developing Relevant Strategic Initiatives

The strategic engineering approach also seeks to consider interventions as solutions to the situation being studied. It consists of basing the intervention on the identification of strategic options that should be deployed to achieve the desired end state (DES) and make the vision a reality by addressing the challenges inherent to the major issues arising from the analysis of the strategic landscape. As an analogy with the context of this landscape, the solution recommended at the global level takes the form of a system of options. Each of these options is itself broken down into suboptions relating to the different fields of intervention; these suboptions are in turn specified in the form of profiles of action mechanisms.

How should we approach the development of strategic options? How can we guarantee their consistency, that is, the cohesion of their components, and ensure that they respond to the problems identified within the strategic landscape?

3.1. Defining objectives and considering options

We can try to develop strategic initiatives based on the vision or major orientations. The vision is the line of force or the fundamental axis that must govern the trade-offs made in determining strategic action.

The approach consists of clarifying the purpose associate with this vision and breaking it down into increasingly detailed and operational objectives and sub-objectives. Efforts are made to outline the operational objectives that contribute to the accomplishment of the concrete vision. For Cartesian thinkers, this is a question of identifying the various strategic "target–objective" actions or initiatives. This consists of defining lines of action or combinations of strategic options to achieve one or more operational objectives. To some extent, this confirms how appropriate they are in relation to the vision.

3.1.1. *Specifying and defining the main orientations*

The process that led to the construction of the exploratory scenarios highlighted the key issues and major uncertainties that could affect the company's environment, as well as the concepts that should be attached to them.

The scenarios themselves are relevant, consistent, and plausible combinations of assumptions about the future. Thus, by immersing oneself in the logic of the scenarios through the specification of their underlying assumptions, it is possible to flesh them out according to the concerns of the company. This work acts to a certain extent to create alerts, which if shared, can make it possible to broaden the spectrum of possibilities and to modify the company's view of the spaces it has for maneuvering in different fields (marketing, technology, distribution, etc.). The prospective exploration thus plays a tacit but decisive role in forging the company's long-term vision, as well as its intentions for strategic action, as anticipatory responses to the key elements of the problem.

As part of a participatory approach intended to clarify the implications of certain scenarios for a given issue, the company strives to provide answers to questions that remain unanswered. Thus, the description of each scenario facilitates the identification of appropriate strategic or marketing recommendations, while taking into account the singularity of the future of the scenario in question. By repeating this exercise for each of the scenarios, multiple "scenario-frame" strategic initiatives are formulated to cover – at least partially – the many possible future outcomes. These initiatives are logically appropriate to the scenarios from which they were developed.

While the above approaches represent a need for rationality, there are still other ways of identifying strategic initiatives. These are "one-off" projects, the logic of which is more subject to the pressure of opinion or corporate interests or the personal will of members of the management team. These proposals, even if they escape the previous logic, cannot be excluded *a priori* from strategic initiatives, which is worthy of an examination.

In observing the way in which strategic initiatives emerge, we cannot ignore the importance of those that have followed an atypical path compared to the normative approach, too often considered the more favorable path forward in this area.

3.1.2. *Drafting consistent action profiles*

Regardless of the genesis of the strategic initiatives envisaged, they must be consistent, which is to say plausible, substantial, and differentiated from each other.

From this point of view, consistency becomes a fundamental property to avoid relying on incoherent, inadequate, or even inconsistent courses of action.

3.1.2.1. *Valuation of useful information around the concept of requirement*

All the information necessary for the strategic foresight process (factors of inertia and change, ruptures, preconceived ideas, aspirations and preoccupations, etc.) benefit from being synthesized in the form of a list of requirements. The term "requirement" is both generic and concrete. It reflects both the internalized form of constraints and obstacles (problems of all kinds) as well as aspirations and ambitions. Demands are both a focus of the prospective approach and a concrete angle of attack for strategic action. They can be expressed in terms of issues and challenges because of their inclination to pose questions for the actors.

The approach, in terms of requirements, allows us to focus from the outset on the salient factors. By its nature, it gives more emphasis on both the reflection and the resulting documents.

3.1.2.2. *Identification of strategic levers among controllable requirements*

To increase the efficiency of the process, the information must be subjected to a rigorous sorting procedure. This cascading procedure is based on three successive key elements that refer to the definition of the future, considered to result from necessity, will, and chance:

– The first sorting key is the *importance of the requirement*, that is, its inescapable character. Initially, only the most important factors or requirements providing the most structure will be taken into account.

– The second sorting key involves *mastering this requirement*, which means the main actor (the company, organization, local authority, etc.) must have the effective ability to maneuver in order to respond to aspirations or remove obstacles. Among the requirements, those deemed manageable will be addressed in the context of the identification of *strategic orientations* and *actions*. On the other hand, the requirements that are not controllable will be used for *prospective exploration*.

– The third sorting key involves *uncertainty*: it reflects the way in which this requirement could manifest itself in the future, or the answer that should be given to it, with the latter remaining undetermined at this stage.

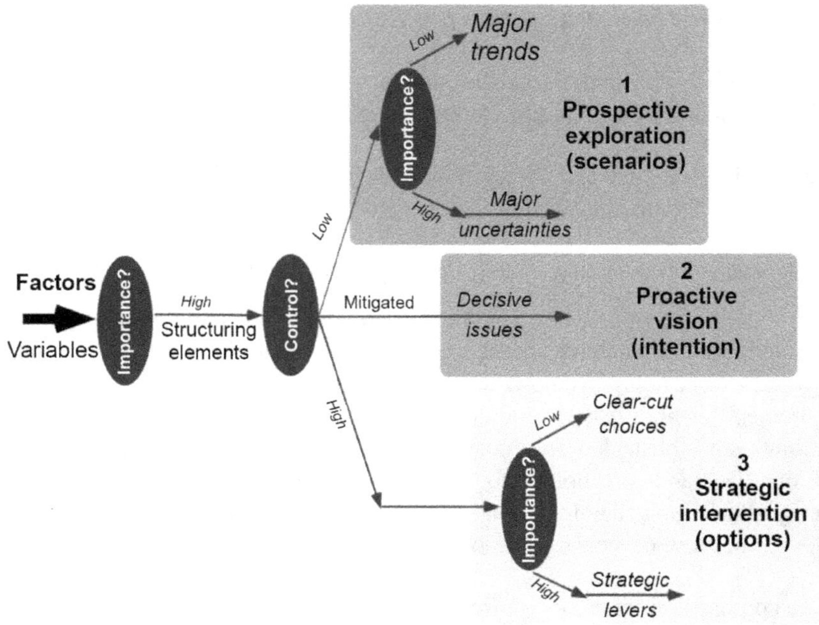

Figure 3.1. *Grid for sorting requirements: between major uncertainties and strategic levers*

The *uncontrollable requirements,* which manifested themselves in ways that are predetermined, correspond to the *major trends*. On the other hand, those which were judged to be the most uncertain (*major uncertainties*) gave rise to the determination of *contrasting hypotheses* on the studied horizon. Exploratory scenarios were also constructed, thanks to morphological analysis, as probable combinations of such hypotheses.

In accordance with the cascade sorting procedure presented in Figure 3.1, the factors and variables that are found to be important and controllable (capable of being controlled, by influencing the course of their evolution) fall within the scope of strategic action. Paradoxically, in this field, where it is possible for willpower to be exercised, the question of uncertainty arises. However, in this case, uncertainty refers more to the notion of non-decision or non-resolution in choices than to the influence of chance, inherent in the field of prospective exploration. As in this field, the variables and factors in the strategic field are fragmented into two subgroups, according to the degree of decision-making uncertainty attached to them:

– *controllable* requirements that have already been the subject of *clear-cut choices* contribute to excess determination in the decision and action;

– the requirements for which answers are still *uncertain* – because the choices have not yet been decided – are the *strategic levers*.

3.1.2.3. *Fabrication of option profiles by morphological analysis*

Although a voluntary and normative approach, the development of courses of action or strategic options at the level of the company, organization, or local authority can be as complex as the construction of scenarios.

In order to guarantee a minimum level of consistency and to try to reduce this complexity, the same approach should be used as for the construction of exploratory scenarios: a morphological analysis.

The three steps of this process are as follows:

1) identification of major challenges or strategic levers that generate decisive arbitrations;

2) identification of optional positions for each lever;

3) development of strategic profiles as combinations of options for each of the different levers.

Interpretations according to domains / Generic steps of the method	**Prospective exploration**		**Strategic inspiration**
	Approach	Outline	
1/ **Essential dimensions**	Major uncertainties or key questions for the future	Major uncertainty 1 Key question 1. Major uncertainty 2. Key question 2. Major uncertainty 3. Key question 3.	Major challenges or decisive levers
2/ **Alternative configurations**	Hypothetical situations or response modalities	Key question 1. Key question 2. Key question 3.	Optional positions
3/ **Contrasting configuration profiles**	Combinations of hypotheses or exploratory scenarios	Key question 1. Key question 2. Key question 3. Scenarios Sc A Sc B Sc C	Conjunctions of positions or profiles/ strategic options

Figure 3.2. *Mechanisms for the use of morphological analysis*

This method of construction of strategic profiles (or frameworks) is not independent of the exploratory scenarios. It can be practiced according to different degrees of consideration of such scenarios. First of all, the strategic levers are determined (step 1), regardless of the degree to which they depend on the scenarios.

The next two steps of the morphological analysis include a possible reference to the scenarios. They concern the previous identification of the options relating to the different decisive levers (step 2). On the other hand, each strategic profile will be constructed as a response chosen by the company or organization in the event that a given scenario is realized. Thus, the options are the most relevant responses with regard to the situation described by the reference scenario.

In order to take into account the exploratory scenarios, it is important to refrain from defining strategic options independently of the scenarios (step 3). In this case, it is a matter of "customizing" them to the company's desirable reactions to the occurrence of the reference scenario.

As with the scenarios, the construction of strategic profiles can be undertaken by following a two-stage procedure:

– the preliminary stage includes the identification of the decisive levers and the subsequent grouping of these into Strategic Intervention Areas (SIAs). These areas cannot exist prior to the identification of the decisive levers and are therefore not generic. They have a unique character, inherent to their applicability to the context and to the problem of the action;

– the first stage involves the construction of strategic profiles for each of the SIAs, using the morphological analysis method (combinations of options);

– the second stage consists of the creation of global strategic profiles; this stage is carried out by aggregating the profiles or composite options obtained for the different SIAs.

3.1.2.4. *Creative imagination allows options to be expanded*

The use of analytical approaches should in no way obscure the originality and innovative nature of the recommended solutions. It is therefore preferable to consider other modes of identification based essentially on intuition and "creative imagination". Furthermore, it is in these ways that original ideas most often emerge and thus they should not be rejected *a priori*. In addition, these approaches can be opportunely linked with morphological analysis, both in semantic research relating to the titles of global options or by domains, and in the discovery or naming of "elementary options" or contrasting modes of action.

3.1.3. *The case of the industrial future of a department in Ile-de-France*

At the end of the 1990s, we were asked to provide methodological support for a prospective study with multiple stakeholders carried out by the Paris Chamber of Commerce and Industry concerning the industrial future of the department of Hauts-de-Seine over the next 12 years. One of the particularities of this reflection consisted of preparing four profiles of the strategic options using the morphological analysis method, in response to the four scenario profiles previously outlined.

In this instance, special attention should be paid to the fact that such a substantial fit between objectives and scenarios is singularly less frequent than one might think, inasmuch as:

– the identification of strategies most often exists prior to the development of scenarios;

– these, when they exist, are already obsolete or too broad in scope to be useful;

– the entities respectively in charge (internally) of foresight and strategy are ill-equipped, too dissimilar, and reluctant to collaborate effectively.

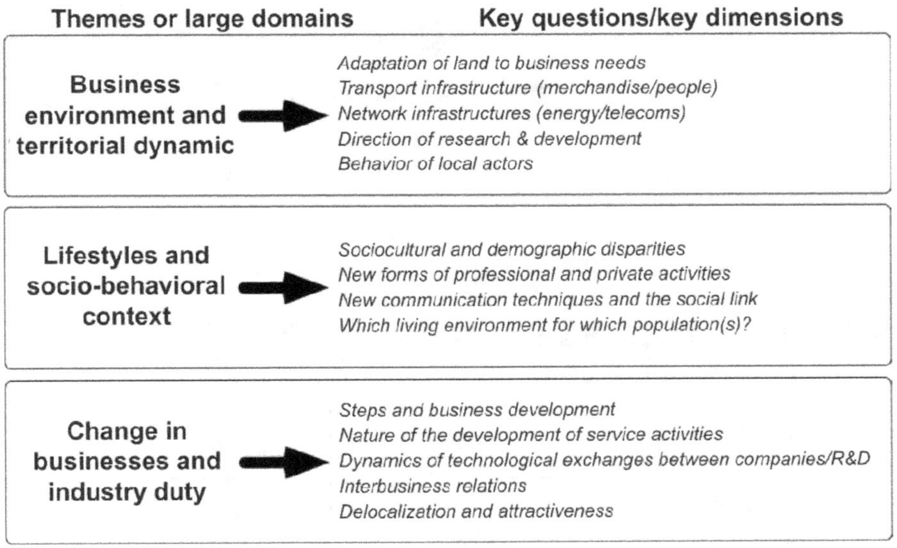

Figure 3.3. *The relevant structuring of the problem: 14 key questions grouped into three themes*

As an illustration, Figures 3.3, 3.4, and 3.5 show the summary results of the emblematic phases of the strategic anticipation process carried out by the Hauts-de-Seine department, namely: the structuring of the problem (Figure 3.3), the construction of exploratory scenarios, noting their conditions of occurrence (Figure 3.4), and the profiles of strategic options that were closely associated with them (Figure 3.5).

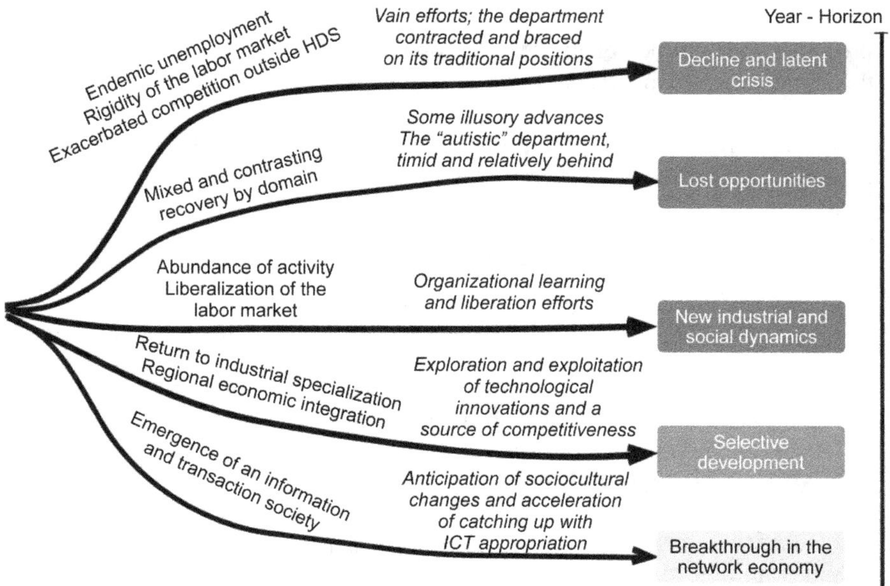

Figure 3.4. *Synopsis of the occurrence of global scenarios*

Framing scenario	Adapted strategic profile
Decline and latent crisis	*Contain the crisis and strive to bounce back*
Lost opportunities	*Reveal and promote real potential*
New industrial and social dynamics	*Promote solidarity and stimulate economic and social synergies*
Selective development	*Focus on and excel in distinctive, innovative extremes with high added value*
Breakthrough in the network economy	*Take a stand in electronic intermediation and knowledge engineering*

Figure 3.5. *A strategic profile for the department for each global scenario*

3.2. Transforming options into project concepts

Identifying coherent and effective strategic options for certain objectives ensures that the recommended solutions are valid. However, this is only of interest if they are translated into concrete actions and put into perspective in relation to the company's vision, the foreshadowing of the future and the requirements of the environment and the sociopolitical context. The actions can only be implemented and monitored if there are indicators to assess and measure the results achieved at the end of the project. Taking the interactions between the objectives of each strategic option into consideration using a tree structure makes it possible to first validate the adopted system of objectives on several levels and complete the information if necessary. Secondly, it is important to determine the critical conditions, to detect possible contradictions between objectives and to determine their importance. The goal of this is to trim down the tree of objectives to ensure satisfactory "feedback" for the effects, or even changes, expected from the strategic option at the global level. Finally, the different options revisited in this way should be presented in an integrated manner, in the form of a logical framework, in order to involve the entire organization concerned.

3.2.1. *Creating the tree of objectives*

The main orientations are expressed in the form of a goal tree. This tree consists of a graphic description of the objectives of a strategic option or project, linked hierarchically through a diagram. This diagram is organized in the form of a tree structure, whose branches represent the means-ends relationships between the different objectives. The most operational objectives are located at the lowest level of the hierarchy. These are the connection points for actions, for them to achieve their higher-level objectives. This presupposes that logical causal links have been identified between the objectives within a given vision, strategy, or business project.

This tree structure helps to define and prioritize project objectives in an explicit and understandable way. The breakdown of objectives into subobjectives highlights the interdependence of the different levels in achieving the higher-level objectives or goals.

3.2.1.1. *Direct construction by presentation of orientations*

This tree is all the more valid and useful if it is prepared in a group with the participation of the people or entities who are the stakeholders in the implementation of these objectives.

There are two variants in the direct graphical construction (without prior analytical processing) of the tree:

1) draw the tree directly as the group develops the tree's branches;

2) write down the objectives respectively, then group and classify them in order to structure and draw the tree.

Box 3.1 describes the procedure to follow in either of these cases.

The process can be summarized in six steps. These, and the main operations involved, are listed below:

1) *Develop a list of objectives*: define the area to which the strategic initiative responds, then identify as many objectives as possible without structuring or prioritizing them, expressing them as: infinitive + verb + qualifiers (for example, "master the use of strategic action analysis tools").

2) *Identify an overall objective*: choose an overarching objective or purpose (more global, longer term, etc.) on which all the others will converge, then place this objective at the top of the hierarchy, at the top of the tree.

3) *Move down one level in the tree*: select the objectives for that lower level from the list of objectives established in step 1; identify, if necessary, other subobjectives required to achieve the higher objectives to complete the list; draw the links (in the form of line segments) between the objectives and subobjectives.

4) *Extend the tree level by level*: choose an objective at the top level and repeat the previous step; repeat the operation for each objective at the top level; then continue to go down the tree by repeating the last two operations.

5) *Revise the tree if necessary*: check to ensure that there are no missing objectives or intermediate levels; if necessary, decide whether the tree should be further extended.

6) *Check to ensure that it is possible to measure the objectives at the lowest level*: for each objective, ask to find out about the necessary indicators and their availability; finally, the tree structure should only stop if the objectives can give rise to concrete and/or measurable actions.

The more operational objectives at the lower level will serve as focal points for activities and allocation of project resources and means.

Box 3.1. *The six-step objective tree*

The tree of objectives is based on the principle that each node is both an objective for those at the lower level attached to it and a solution or a support (subobjective) for those at the higher level to which it is attached.

This type of tree structure offers the possibility of exploring different angles of attack to achieve all of the project's objectives and makes it possible to evaluate the degree to which the means implemented to achieve them are relevant.

3.2.1.2. *Other complex modes of prioritizing objectives*

Sometimes the structure of objectives within a complex strategy or project is not obvious and cannot be directly broken down into increasingly precise levels. In this case, an attempt can be made to approach the logic behind the objectives by systematically reviewing them. As with a structural analysis, it is important to first develop the table describing the logical relationships between objectives. More specifically, this is a matrix of cross-impacts that is a binary reflection of whether or not there are relationships that directly support the objectives on each other.

3.2.1.3. *Multilevel incorporation of interactions of objectives*

This step consists of qualifying and then quantifying the interactions between the impacts of each policy option, according to their impacts at the levels of the overarching levels of the objectives tree. If necessary, the impact information for that level is modified as appropriate. The interactions are then semantically synthesized in terms of their contribution to the higher-level objectives.

3.2.2. *Detecting the critical conditions for achieving objectives*

In order to determine the critical conditions, it is necessary to proceed in stages:

– the first stage is to define the prerequisites to avoid contradictions between objectives. To do this, the salient contradictions between objectives should be identified at each outcome level. Next, the threshold of acceptability of these contradictions must be defined. Finally, the contradictory objectives will be cut out;

– the second stage involves the decision of whether or not to validate the withdrawal of certain critical objectives by requiring that their pruning is not detrimental to the whole;

– the third stage aims to define the vulnerabilities inherent to the context, in order to prevent risks. In order to do so, it is imperative to identify the vulnerabilities related to the negative impacts or unacceptability of the actions that make up the strategic option.

The goal is to subsequently examine the existing actions or measures that could weaken or even call into question the strategic option in question. It is also necessary to set out realistic proposals ensuring that these vulnerabilities will not be exacerbated.

Next, the incorporation of complementary actions or measures to accompany the strategic option will help to reduce critical vulnerabilities. Finally, some of the results of the strategic option could, if necessary, be revised and placed at lower levels.

The connection between the hierarchical structure of the problem studied and the associated objective tree guarantees a more coherent and synthetic formulation of any solution-project in the form of a logical framework.

In particular, the existence of a generic graph, allowing for problem factors to be associated with objective-solutions at each level, has the following advantages:

– it gives a view of all the possibilities of progress that allow reaching the most global and fundamental objective;

– it widens the range of angles of attack of the problem and makes it possible to generate more alternative project solutions;

– it makes it possible to constitute a grid for comparative evaluation of the relevance of all projects in relation to the same overall objective; hence its name, "relevance tree".

The formulation of the intervention project in the form of a logical framework brings together all these aspects. It encompasses the actions that make up the intervention project (blackened tops) and the objectives at different levels of their chain of effects, which form a subchart of the generic objectives tree.

This concept of formulation is subject to the following condition: these impacts will only help to achieve the fundamental objective *in the end* if the requirements or problems underlying the objectives that are not affected by the envisaged actions do not become worse. Indeed, the context in which the project is inserted must not become an obstacle to its achievement. It is around these latter requirements that we will identify the corpus of framework hypotheses which, although not controllable at the level of the project, are nevertheless conditions for its success. This corpus will constitute an initial platform of *important presuppositions* for the project.

Breakdown of the problem/key requirement
Objectives/relevance tree

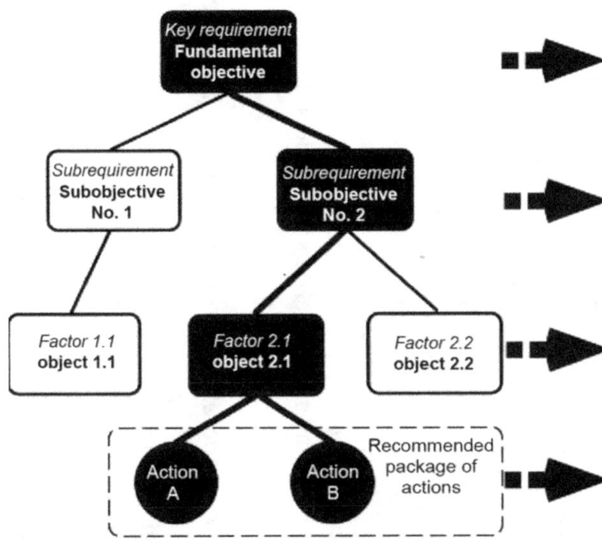

Logical framework for project formulation

Chain of objectives Results-actions	Indica-tors	Framework assumptions Important presuppositions
Fundamental objective: Satisfy key requirement		Realization of the framework scenario (reference) of the course of action
Goal: Subobjective No. 2 Satisfy key requirement No. 2		Nonexacerbation of the subrequirement
Results: Object 2.1: Explicative factor 2.1 evolves favorably		Neutralization of the explicative factor Nonaggravation of vulnerabilities No disapperance of assets
Track / body of actions: – *Action A* – *Action B*		Other framework conditions likely to guarantee the transition from the body of actions to results

Figure 3.6. *From the project-solution to its logical formulation*

These presuppositions, organized hierarchically into levels, may be supplemented by hypotheses resulting from:

– an examination of the long-term validity of the project and in particular the non-existence of environmental scenarios with the potential to generate significant counter-performances, which may even be prohibitive to its success;

– the confrontation of the intervention with its context, and in particular with its "sociopolitical landscape". In particular, the presuppositions concerning the concretization of assets (advantages and major supports) or the control of its vulnerabilities (disadvantages and critical resistance);

– the possible contradictions perceived by the actors or detected with regard to the different contexts (technological, economic, ecological) as they emerge from the analysis of the convergences/divergences between actions or objectives. For example, a project whose actions or objectives are deemed to be coherent from the point of view of the strategic planner could be controversial, to the point of creating a coalition of detractors. A key dimension of the sociopolitical analysis of the project offers the possibility of preventing such situations.

The generic objectives/relevance tree allows us to establish the necessary correspondence between the explicit objectives of a project and its conditions of success. Ignoring this dialectic amounts to privileging the orientation of the project to the detriment of its realism. This leads us to reflect on the statements of some crisis specialists, who say that the failure of too many actions reflects the resurgence, or even revenge, of problems that have been eliminated and objectives that have been "forgotten".

Finally, one of the criteria for the validity of a project is that its direct manager (the project leader) should be able to assess the accomplishment of the expected results, as long as he or she has sufficient resources. This requirement means that there must be no framework conditions attached to the project activities (at the lower level of the objective tree) that could jeopardize the achievement of its immediate results. Otherwise, the project will have to be revised, either by adding appropriate actions or by accepting expected effects that are less ambitious.

3.2.3. *Summarizing each option in the form of a logical framework*

The "logical framework" is a set of intersecting concepts that can be used to *formulate an intervention project* explicitly and *facilitate its evaluation*. It seeks to identify and clarify the logical links between the objectives of the intervention, the actions it involves, and its results. It deals with these links in the form of proposals for a hierarchy, in terms of causality or a chain of effects. It provides a means of highlighting and concretely describing uncertainties in the project. In addition, it provides a means to make explicit the external factors that may influence the implementation of the intervention, and possibly the likelihood of its success.

Once the options are reviewed one by one, they can be presented in an integrated fashion to reveal the coherence of the whole.

To do so, a logical framework for the formulation and contextualized monitoring of the recommended option should be developed. This should be presented in the form of a summary sheet including the following information: the main background (origin of the idea, context of intervention, foundation of the problem), the summary description (main goals and objectives, action program and expected results, means to be used), the elements of justification (appropriateness to the strategic frame of reference, ability to achieve the fundamental goal, validity with regard to future scenarios, capacity to integrate into the context), and the conditions for implementation and success (underlying hypotheses, prerequisites, requirements and presuppositions, main accompanying measures required, types of monitoring envisioned and evaluation devices).

The logical framework method, which applies to both tangible and intangible projects, appears to be well suited to this problem.

The essence of the use of the concepts underlying this logical framework can be summarized in a four-row, four-column, two-entry square table (4 × 4 square matrix).

The logical framework allows us to:

– *Summarize, on a single sheet,* in a simple, clear and relatively complete way, the most important elements of any intervention project, whether tangible (equipment, investments, etc.) or intangible (training and capacity building, study);

– *Summarize the basic elements of the project design* with the goal of either refining them *ex ante* or controlling them for monitoring-evaluation.

Using the logical framework, the managers and executives involved are given a *synoptic vision* of the project without having to refer to large volumes of documentation to identify objectives, specify and measure indicators, or evaluate external success factors.

This logical framework can be used in three different and complementary ways, discussed in the following sections: as a tool for assessing and communicating about the project, as a means of planning it more comprehensively, and as a reference for monitoring and revision.

3.2.3.1. *A tool for appropriation and communication on the project*

The logical framework should be developed collectively with the participation of those responsible for the design, programming, and implementation of the project. This offers a double opportunity to reach a consensus on the objectives and the means to monitor progress and possible changes. As such, the preparation of the logical framework should not be limited to a simple exercise of filling in the empty boxes on the table, as if filling out a form. It should serve as a catalyst and facilitate good collaboration between most of the people and/or entities that – at different levels, or at different times – are involved in the process of developing and/or implementing the project. As a communication medium, the logical framework provides a synoptic version of the project to decision-makers and everyone who will be in charge of managing it.

3.2.3.2. *A method for more effective planning of the intervention*

The logical framework is also a tool for the formulation and planning of the project, insofar as it offers managers:

– an excellent guide for stating the objectives of the project, the expected results (for which it is directly responsible), the activities and means necessary for its implementation and operation, as well as the factors beyond its control that constitute the conditions for obtaining the results;

– a means of clarifying its responsibilities. By identifying the external factors that may affect the possibility of achieving the project's goal, the logical framework has the advantage of delineating the project manager's area of responsibility and freeing him or her from any conditions within the framework that are deemed beyond his or her control.

The project manager cannot take full responsibility for the success of the project. The assumptions made about the realization of contextual conditions that depend on the prerogatives of the hierarchical level or related entities play a decisive role. However, the project leader must ensure that the expected results are achieved, and that the activities are carried out. The means must be provided and used according to the standards set by the quantitative indicators, but especially the qualitative indicators (results and deadlines, etc.).

3.2.3.3. *A reference for feedback and review*

The logical framework also serves as the basis for the development of a monitoring and evaluation grid for the project. At each level of the chain of effects (activities, outputs, purpose, objectives, etc.) and at different points in the project cycle, different indicators provide the measures that can be used as a starting point for an assessment. These indicators are accompanied by a list of concrete mechanisms for verification.

However, depending on the context and timing of the evaluation, these indicators will need to be supplemented with more elaborate questions made in light of the experiences or major developments that occur during the implementation of the project.

The logical framework for formulating a project is a square table (4×4), developed in the 1970s through the initiative of the US Agency for International Development, and whose use is currently recommended by various agencies and international cooperation institutions. The concept of this logical framework is based on the fact that a project is a set of actions and mechanisms chosen among various options to achieve a goal, which is a specific objective, and must therefore be specified. The project is an experiment based on the hypotheses of cause and effect. Therefore, it is a learning system that will show how, and to what extent, the results are achieved. The project is only a subsystem that must be able to fit into a broader context, including organizations and institutions, other projects, and various actors whose interests do not necessarily coincide with those of the project. This context creates uncertainties about the success of the project that need to be identified at the outset.

1) The *logic behind the preparation of the framework* is twofold, facilitating, in a practical form, the planning and evaluation of the project:

– The *vertical logic* reflects the causal links that characterize the project. It provides the direction of the chain of propagation of project impacts. It provides the justification of the project from the point of view of its relevance to its explicit objective (bottom-up), as well as the progressive concrete effectuation of this objective in the form of concrete actions (top-down). Following this logic, the different levels are specified: fundamental objective, specific objective, the expected results, and the means deployed;

– The *horizontal logic* seeks to identify what is to be produced or achieved, as well as to describe how to verify that it is actually achieved. It specifies, for each level of impact (tier), the objectively verifiable indicators, the concrete means by which they can be measured, and the framework conditions discussed above.

2) The *process of preparing the framework* should normally follow steps that correspond to the tiers of cells in the matrix. Although one can start anywhere in the matrix, the task is easier if one first looks at the purpose of the project, that is, its specific objective. An indicative version of the sequence of steps is described below:

– *Define the purpose of the project*: the purpose of the project is its reason to exist, which is to say, the justification for pursuing results. In analytical terms, the goal is the proposed solution to a specific question or problem. It is the desired effect: what the project will change as a result of its results. The goal is often at the level of behavioral change among its beneficiaries. In principle, and for practical reasons, a project should have very few goals; it is even preferable to have only one.

– *Specify the results that lead to the goal*: the results are the "outputs" (or "ex-outcomes") of the transformation of the system that the project constitutes. In other words, they are what the project will accomplish. They are the contractual commitments (specifications or terms of reference) of the project. They therefore reflect the elements of the situation that can be expected at the end of the project and which require resources to be mobilized.

– *Identify the key activities that lead to results*: it is imperative to specify the project activities in order to highlight the cause-and-effect relationships that describe the processes that produce results. The project's activities should not be included with the resources and factors (building materials, labor, vehicles, financial means, etc.) needed to carry them out. These resources act as the "raw materials" of the project. They are consumed by the project's activities.

– *Set the overall objective and ensure the vertical logic of the project*: after describing the project's goal–outcome–activity levels, we can move on to the next level: that of the overall objective. This is the broadest objective to which the project should contribute. It represents the goal of a broader program and therefore that of a portfolio of actions or interventions to which our project belongs. It can also correspond to the objective through which the project is linked to the strategic vision. The vertical logic of the project includes the *hierarchical breakdown of* the problem that the project aims to solve (or the question it is supposed to answer) by highlighting the proposed solution and the actions to be taken to achieve it. The linkages between the four levels of the logical framework are expressed in terms of necessary hypotheses, that is, sentences in the form of "if... then" that reflect the causal sequence of project impacts that allow the project analyst or manager to easily visualize the internal structure of the project in terms of its validity and logic.

– *Identify the presuppositions related to each level*: the presuppositions, which form the assumptions about the framework conditions, play a very important role in the development of the logical framework and in defining the mission of the project manager or project leader. By identifying the contextual conditions that may influence the project's results, the presuppositions help to identify circumstances that are beyond the control of the project management, but which must be taken into account. They may also mention other projects whose results may catalyze those of the project under consideration. Sometimes it is desirable to lower the ambitions of the project, or even to abandon it when the conditions defined by the important presuppositions seem so unlikely as to call into question the existence of the project.

– *Identify valid indicators for each level of the logical framework*: verifiable indicators are measures that show what the project has accomplished. They are a very important aspect of the logical framework and must verify certain properties (being plausible, reflecting goals or targets, representing the level of generality associated with the level in question, etc.). They should be expressed in terms of quantity, quality, and time frame, so that two people can judge the success of the project, even if they have different perspectives on the project.

– *Specify the means for verification associated with each indicator at each level*: the details in this third column of the logical framework make it possible, for a given level, to, determine whether the indicators are available or whether specific measures for obtaining them should be included in the project's actions or means. This also involves determining whether the indicators are "sustainable" and, if possible, how to ensure that they are so.

– *Periodically review the logical framework in the light of experience*: to be effective, all management must be flexible and include detailed changes in project implementation. To prevent the logical framework from becoming an academic exercise that quickly becomes outdated, the project team should review it periodically to ensure its continued validity and make any necessary corrections. In this case, the logical framework becomes a tool for strategic oversight and monitoring – an evaluation of the overall course of action taken for the project.

Box 3.2. *The logical framework method: usage instructions (University of Pittsburgh, USA)*

3.2.4. *Aggregating projects into coherent programs*

The preceding considerations lead us to take an original approach to the conceptual transformation mechanism of projects into action programs: a mechanics that consists of accessing a higher order of general applicability (or metalevel). In reality, this is a concern for many institutions, including at the international level (the UNDP, the World Bank, etc.), and a major challenge for the strengthening of development policies.

From a conceptual point of view, a program is at a higher level of general applicability and complexity than intervention projects. This implies the consideration of new, relatively subtle requirements and additional conditions. However, the formulation of a program calls for similar notions. From this point of view, it is conceptually possible to synthesize any program, as with projects, in the form of a logical framework: a chain of objectives and conditions.

3.2.4.1. *Comparison of project and program objective levels*

The project's and the program's logical frameworks are shifted up one level, so that from the perspective of the chain of objectives:

– the results of the project correspond to the actions of the program;

– the specific objectives of the project correspond to the results of the program;

– the development objective of the project corresponds to the specific objectives of the program.

Moreover, the level of the project actions does not correspond to the program level, as it is at a level of precision that cannot concern that level.

In addition to these connections in terms of the content of the levels between projects and programs, the question arises of the principles of aggregation that make it possible to move from the broad objectives of the projects to those of the program, and then from the framework conditions of the various projects to those of the program.

At each level, the program's objectives are achieved by pooling and consolidating the specific objectives of the various projects involved. For example, the actions of the program include the results of the different projects of which it is comprised.

Likewise, the results of the program are obtained through the consolidation (qualitative and quantitative) of the specific objectives of the different projects it is supposed to carry out.

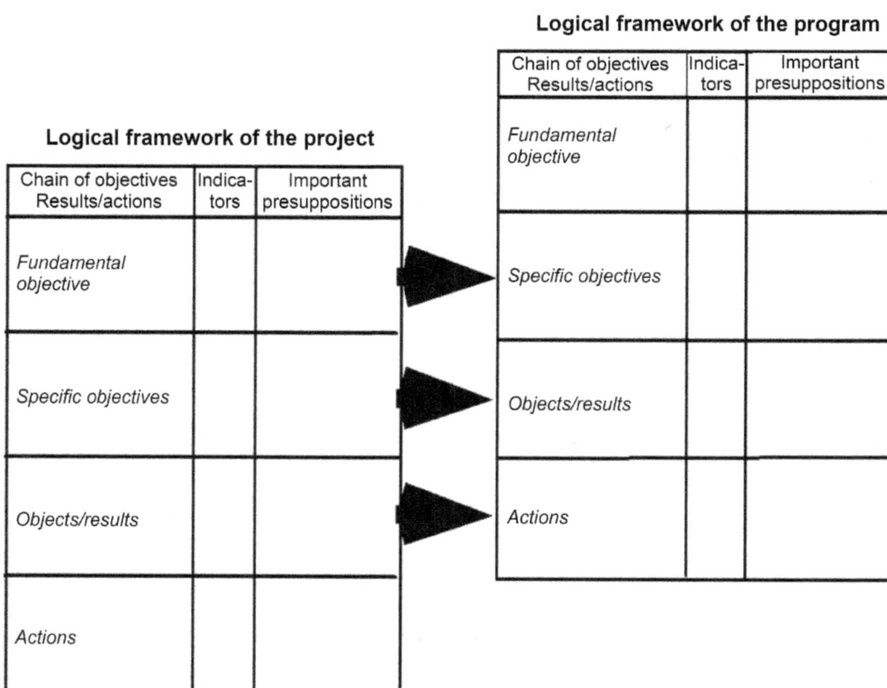

Figure 3.7. *Correspondence between project and program objective levels*

With regard to the determination of the framework conditions of the program, the principle of aggregation, for each horizontal layer, consists of relating the environmental assumptions (important presuppositions) of a given project to the equally ranked objectives of all other projects. The purpose of this exercise is to identify the reciprocal compensations that could result. In this way, the cross-cancellation of the conditions for the implementation of certain projects with equally ranked objectives from other related projects can lead to the satisfaction, and therefore the elimination, of some framework conditions specifically attached to the projects at the program level.

However, beyond these mechanistic-type correspondences, this program must verify certain fundamental properties in order to be coherent and relevant.

3.2.4.2. *Reasoned inclusion of accompanying and support measures*

All of these objectives must be able to converge on the same higher-order goal, which is the fundamental objective of the program. This requirement, however obvious it may be, can lead to the rejection of projects that are not congruent with the higher-order objective identified.

Moreover, the validity of such a program – as an encompassing project (or meta-project) – depends on the verification of the same criteria of coherence as those imposed on the different projects of which it is comprised.

In particular, it is imperative that the framework conditions at the level of the program's activities cancel each other out, so that the achievement of its immediate results can depend solely on the responsibility of its coordinator.

This last requirement requires, in most cases:

– either adding other actions to the program, thus adding other projects in particular;

– or resorting to the implementation of accompanying or support measures (macroeconomic, regulatory, lobbying, etc.), with the goal of loosening the bottlenecks related to the constraining conditions and which may affect the achievement of the expected results of the program.

If we refer to the experience of strategic planning accumulated in companies as well as at the level of sectors of activity or countries, this leads us to the observation that the first condition for the convergence of the objectives, while the easiest, is only rarely respected. How many plans of institutions, organizations, or companies have included programs that are essentially limited to juxtapositions of projects or catalogs of measures? Indeed, the real common denominator is that they belong to a given field or sector, are institutionally anchored in the same entity, or are mainly funded by a particular major donor.

The second condition, which appears more subtle, shows that a set of projects or measures, even if they converge towards a goal that transcends them, is not enough to constitute a true program. It requires a *critical mass* of interventions in the broadest sense, that is, a sufficiently consistent grouping of projects or policy measures, to ensure that the problems are mutually *overcome* and that the program is sustainable. Of course, meeting this last requirement requires many adjustments to the original program. These relate not only to the package of projects and accompanying measures concerned but also to the content of each of them. It is therefore understandable that it can be very simplistic to confine a real program (that deserves to be addressed as such) to projects or measures carried out by an appropriate entity (departments, ministries, subsidiaries, institutions) or financed via a particular budget line or source.

Thus, it is difficult to implement development programs that are not simple amalgamations of projects but can claim to be innovative and capable of "holding their own".

3.2.5. *Considering the plan as a metaprogram of actions*

Moving the reasoning put forward in section 3.2.4.2 to a higher level, a plan can similarly be defined as a metaprogram of interventions.

3.2.5.1. *Synergy between action programs and support measures*

To this end, a plan must include the measures and actions necessary for it to be viable. This implies that it cannot be conceived without consideration for economic policies. On the contrary, it must integrate them in order to be strengthened. Thus, the lack of connections between a plan's investment programs and economic or sectoral reform programs is a sufficient reason for the plan to be deemed inconsistent and unsustainable. However, the institutional anchoring of these two major aspects of managerial action, the origin of their respective support at the level of international bodies, and the excessive opposition between the short term and the long term are likely to accentuate their unfortunate separation and/or overlap.

3.2.5.2. *Identification of interventions to influence goal drift*

There are few companies or organizations that are in a position to develop and conduct large-scale projects or measures without interference from other interventions. For this reason, from the outset, the concept of conducting a comprehensive set of measures and projects would seem to be the most appropriate, if not the only one that would be applicable. It is therefore unacceptable to approach any intervention from the outset without taking into account those that are already being implemented. It is a requirement which, if not met, will become an issue again later, possibly through overlapping. This is why the pragmatic implementation of the logical framework method in a program of interventions/actions in progress calls for a lateral entry into the "strategist's path". The first step for this is to rethink and then transcribe, in the form of a logical framework, the projects currently being implemented. The second step is to identify new action programs that have the possibility of stimulating desirable and relevant changes, while at the same time hybridizing with existing ones. The stages of such an approach, which are more bottom-up than top-down, are listed in Box 3.3 for guidance. In the symmetry of possible futures, the role of strategists is to define the paths that can be taken to bring ambitions and intended purposes to life. But none of their choices are neutral with respect to the others and the many possible futures, for which, more than ever, the "butterfly effect" phenomenon applies.

– Training and development in project design and management (objective trees, logical frameworks, etc.).

– Identification of projects and creation of files on them (project document, activity reports, evaluation, etc.).

– Classification of these projects according to different sorting criteria (institutional anchoring, field of intervention, budget lines, etc.).

– Participatory reconstruction of the trees of objectives of the different projects and transcription of these into a logical framework (to be completed).

– Examination of the interactions between these projects in terms of common objectives, joint implementation, or synergies of means; identification of clusters of projects, if any.

– A summary of the objectives emanating from the projects; enumeration of the major orientations emerging from this universe.

– Assessment of the relative contribution of each project (or program of projects) to the achievement of the various objectives (in terms of direct contribution or budgetary effort).

– Revelation of the hierarchy of the major orientations (relevance tree) detected, through the profile of the contributions from the universe of projects.

– Confrontation of this hierarchy with priorities and recommendations for convergence between the project pipeline and deliberate strategy.

Box 3.3. *List bringing to light the emerging strategies from projects*

PART 2

Judging and Arbitrating: Evaluating and Reconfiguring Recommendations to Decide on a Successful Intervention

4

Defining Priorities and Gauging Courses of Action

An analysis of options based on multiple criteria is an essential and valuable step in the politico-strategic decision-making process. It allows the different options to be assessed in a contextualized and circumstantial manner, providing a comparative assessment of their relevance to the components of the strategic landscape and their contribution to tackling a difficult situation.

Although this type of assessment is often made in companies and is considered to provide enlightening insights for the decision, it is still not sufficient for settling on a *final* recommendation on the intervention to be deployed. At this stage of the strategic engineering process, the presumed validity and consistency of the options considered are not sufficient to guarantee compliance with either the constraints weighing on the situation or the implicit rules prevailing in the strategic landscape.

The evaluation is carried out against a reference framework based on strategic landscape modeling. This chapter describes the principles and content of this specific evaluation framework, as well as the main results it provides.

4.1. Clarifying preferences and comparing options

The strategic landscape is made up of a global systemic framework composed of three spaces or subsystems: the strategic reference (SR), which includes the long-term orientations and vision, the general environment scenarios (GES), and the global context of intervention (GCI).

The strategic intervention options are represented by individual systems incorporating various components. These systems are given in the form of component trees.

A multicriteria assessment of strategic options should begin by integrating the "strategic landscape" system and the "strategic interventions" system, as illustrated in Figure 4.1. This "metasystem" brings the interactions of its two subsystems to light. The interactions of these systems result from the systematic confrontation of the different strategic intervention options with the three subsystems that make up the strategic landscape (general context of intervention, general environment scenarios, and strategic reference).

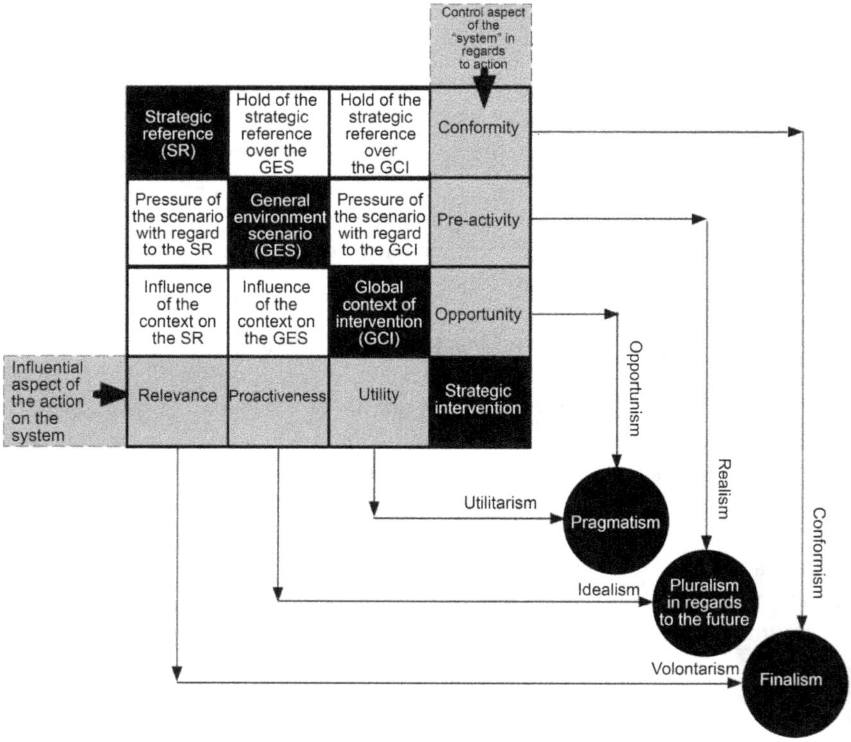

Figure 4.1. *Integrated intervention system: contextualized analysis device for the overall performance of the action*

The interactions are always expressed from two perspectives depending on the direction of the influences under consideration. These are both complementary and contradictory, involving, respectively, the influence of the subsystems on the intervention, and the impact of the intervention on the subsystems. At this global level of the tree structure, the interactions form the basis for the multicriteria evaluation of the overall performance of the different strategic interventions.

4.1.1. *Setting up the overall performance appraisal system*

The first step in the construction of the repository necessary for the evaluation of each strategic intervention is the creation of a multicriteria tree structure.

4.1.1.1. *Determination of the macrocriteria associated with landscape components*

The evaluation is conducted by aggregating assessments of three macrocriteria associated with the different strategic intervention registration registers (see Figure 4.1). These macrocriteria form the basis of the so-called "multicriteria" assessment of the overall performance of the different strategic interventions:

– *Pragmatism* refers to the global context of intervention, which includes two criteria: utilitarianism and opportunism;

– *Pluralism* with regard to the future, or the long-term validity of the intervention, refers to the general environment scenarios and is broken into two criteria: idealism and realism;

– *Finalism* refers to the adequacy of the intervention with regard to the strategic reference and is broken into two criteria according to the aspect under consideration: voluntarism and conformism.

The relationships of the criteria within each macrocriterion are dialectical in nature, as they are both complementary and contradictory, insofar as they take part in a dynamic of mutual fertilization.

4.1.1.2. *Scheme for benchmarking integrated strategic options*

Each criterion can be broken down into finer modules (subcriteria) whose structure is similar to the initial system. For example, the global context of intervention can be broken down into technical/economic, legal/regulatory, and politico-strategic contexts. The last of these contexts can in turn be broken down on the basis of the actors involved, and so on.

The nested frame breakdown method shows a certain similarity between the different levels. However, the level of breakdown or how fine the analysis is done is not the same for all subcriteria.

Thus, the representation of the system is carried out with different depths of the field levels and densities of interconnections, on the basis of the subcriteria.

Like each subsystem which make it up, the broader system of intervention is in the form of a tree of criteria. The deepening of the analysis of certain elements or subelements of the system will result in additional branches of this tree.

We can associate the rating scales that allow us to make local quotations with the thinnest ends of the tree.

The application of this procedure to each element gives our integrated and fractal system its "object-oriented approach" property.

This aggregation is done in a bottom-up manner at different levels to result in an assessment of the overall performance of the strategic intervention, with respect to the integrated intervention system as a whole.

4.1.1.3. *Evaluation of the overall performance of the different options*

The concept of benchmarking the overall performance of policy options takes on a particular meaning, depending on whether the interventions are characterized as:

– alternatives, exclusive of each other, not subject to particular constraints;

– the elementary components of which it is a question of recomposing as well as possible, by taking into account constraints and explicit rules.

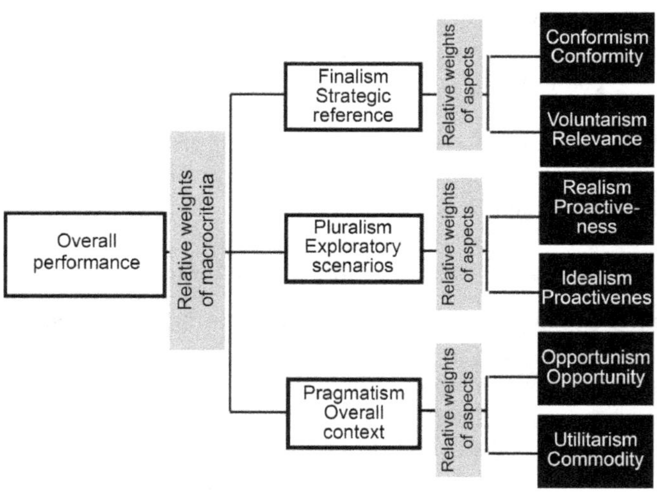

Figure 4.2. *Schematic for multicriteria evaluation design; the overall performance of the strategic intervention*

The evaluation therefore requires a (computerized) differentiated processing according to these cases.

4.1.2. *Structuring and weighting criteria according to their hierarchical breakdown*

Each of the three main dimensions of the macrocriteria must in turn be broken down – or rather interpreted – in the form of increasingly operational criteria for the evaluation of strategic interventions.

4.1.2.1. *Transposition of trees into the form of nested frames*

There is a correspondence between the nested frames, in this case the nested matrices of representation of the contexts of action, and the hierarchy of criteria that they translate. Box 4.1 explains and illustrates this statement.

From hierarchy to contour lines: a methodological impasse

A hierarchical structure presented in the form of a tree (see figure below) refers to the notion of an organization chart and a pyramid-shaped organization. In this model, the top of the hierarchy includes the leaders, whose goals and decisions influence and guide functions at lower levels. Conversely, the latter functions contribute to those carried out at higher levels.

Hierarchical structure of the contour lines

Anti-systemic and counter-productive correspondence

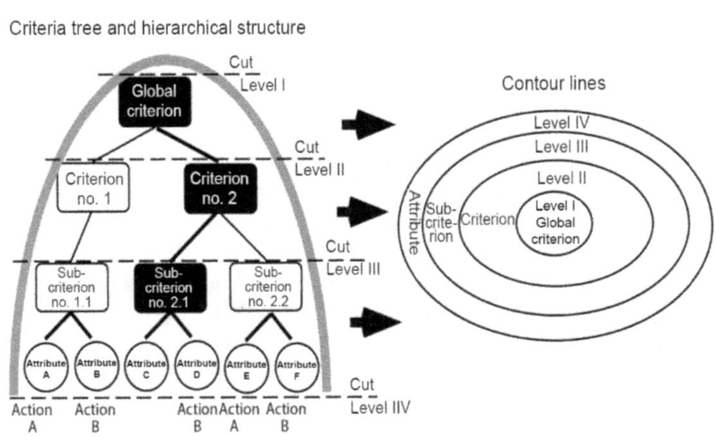

This implicit model is all the more striking because it appears to reflect the tightening observed at the top of the figure, due to the smaller number of higher objectives and vice versa, hence the atomized nature of concrete actions and thus the broadening of the base.

This hierarchy reflects a certain gradation in the variety of contents because the orders from the top are fewer and relatively fixed, while the breakdown of these at the lower levels leads to their splitting into multiple components, the concrete application of which may give rise to numerous interpretations and/or adjustments. The most common conceptual representation of this pyramid and its hierarchy – or its depth – refers to the notion of contour lines: the different levels correspond to nested rings, which become tighter closer to the top.

As attractive as this representation may be, it does not reflect all the elements of the hierarchy and remains unsuitable for a true systemic approach to this structure. Another way of representing this structure is to reverse the direction of the nesting of the frames, considering them all the more encompassing as we get closer to the top and vice versa.

Equivalence between tree structure and nested frames (Ashby 1963)

This new representation is based on the principle that elements at any level of this hierarchy are included in the higher elements to which they are directly related and give them meaning and condition.

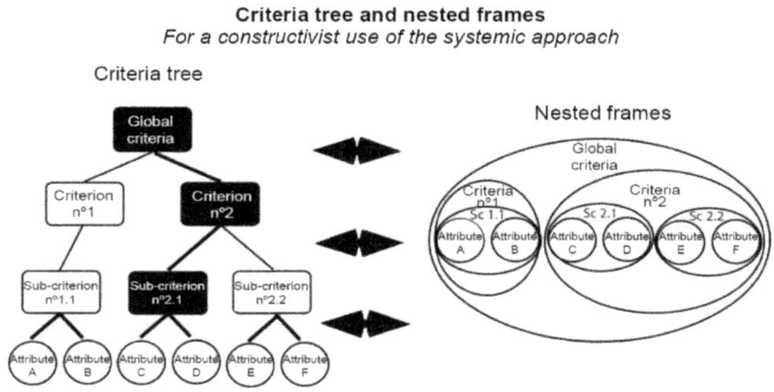

The higher level of the hierarchy here is more overarching and holistic. It serves as a frame of reference for the elements of the lower levels, which are "subsumed" within the higher elements and contribute to making them more concrete by specifying them.

This representation is more in line with that of systemic analysis, in which, according to Ashby, going up one level is equivalent to placing oneself at a higher order of complexity, of greater variety: at this level, the possible states of the system prove to be more diversified. Thus, the latter representation appears to be somehow inverted with respect to that of the contours. Indeed, rather than being limited to representing the horizontal sections of the relief (the pyramid) on which one is at a given level, it is more appropriate to be interested in the extent of the field of vision offered by this level of the landscape. Finally, this representation has the advantage of ensuring an almost automatic and above all one-to-one transposition from a hierarchical tree structure (tree) to a set of nested frames, and vice versa. It is this equivalence that will be systematically used in strategic engineering.

Box 4.1. *Nested frames and tree structure*

4.1.2.2. *Cascading weighting of criteria along tree branches*

The second step in the construction of the strategic intervention options evaluation reference frame concerns the weighting of the criteria of the tree structure.

Indeed, it is appropriate to give weightings to the different nodes of the tree structure in order to calculate an overall evaluation score for each strategic initiative. Establishing these weightings is all the more delicate given that the number of operational criteria is high and that they refer to distinct domains. The tree structure of these criteria can be used to deploy a “cascading” weighting procedure, starting with the objectives. This method is based on the principle that each node in the tree is both a macrocriterion, for those lower-level nodes attached to it, and a subcriterion, for those higher-level nodes to which it is attached.

The method consists, starting from each criterion node of the tree, of deriving the subcriteria that directly depend on it. These weights more precisely reflect the relative contributions of these subcriteria to the consideration of the directly superior criterion.

4.1.2.3. *Distribution of weightings or relative weightings by macrocriteria*

Once the local weightings have been made, the overall weighting of the operational subobjectives is carried out. To do this, it is important not to rely solely on the nodes to which the subobjectives relate but to ultimately refer to the most encompassing, or superior, evaluation criterion. This new weighting is obtained by multiplying all the local weightings (or weighting nodes), located along the branches that connect them, to the higher criterion, which is the root of the tree. This type of tree offers the possibility of exploring different angles of attack to impact the higher order criterion, but also of evaluating their degree of adequacy in relation to it, or their contribution to its achievement.

The weighting comes from the thinking of an expert (or a group) on the issue raised. Without an appropriate tool, in most cases, this thought is difficult to formalize and remains relatively pifometric. Thus, a quick assignment of coefficients will not accurately reflect the reality, unless the problem has already been “mathematized” or the criteria are representative of quantities, which is not relevant here. Strategic reasoning, although it may be based on mathematical representations – because of the intangible nature of the criteria from which it is formed – is formed from heuristic bases.

Software devices facilitate the assignment of relative weights, in a visual and interactive way, based on the expression of the perceptions of the experts involved, without any numerical intervention. The relative weight (or weighting) is then entered into the tree model via a procedure for the revelation and acquisition of expertise. An appropriate computer processing, on the basis of these relative weights, can perform the calculation of the aggregated weights of the nodes of the tree.

The aggregate (overall) weight, as defined above, indicates the share of the considered criterion node in explaining the multicriteria score achieved by a given action.

Box 4.2. *Assigning intuitive weights*

4.1.3. *Determining the estimators and scales of appreciation of the options according to the operational criteria*

The most operational criteria at the lower level should be used, where appropriate, as application points for interventions or policy options. These criteria represent the partial, duly weighted views according to which these interventions will be assessed, through estimators based on attribute and/or rating scales.

The elements involved in the scoring procedure – of which they are an indispensable factor – are the *estimator scales*. Scales of estimators are evaluation parameters in direct relation to the criteria to which they relate and are intended to qualify them judiciously, in terms of assessment or performance. The appreciation can be perceptual (for instance heavy, cold, acceptable) and/or physical (€70, 82%, 500 kg, 2.4 °C). The estimators forming the scales will also be relativized to reflect an expert opinion by means of a weighting from the most to the least significant. The relative weight of the value retained among the estimators during the scoring phase, combined with the aggregate weight of the evaluation criterion, forms the contribution to the final score of the solution/action evaluated.

Box 4.3. *Estimators and rating scales*
(Paul Versteven, Decision Systems Consultant)

4.2. Evaluating options according to the macrocriteria to inform the decision

The pragmatism or operationality of the strategic intervention is based on the assessment of how congruent it is with the overall environment. The idea is to broaden the perspective by representing a metasystem – the “concrete action system” – which encompasses both the strategic intervention to be evaluated and the overall context in which it is inserted. Thus, in accordance with the mode of systemic representation presented earlier, the strategic intervention and the context constitute the main subsystems corresponding to the diagonal blocks in the modeling of the global framework.

4.2.1. *Appreciating congruence with the context of intervention as a springboard for pragmatism*

The guiding principle of congruence assessment is based on a multicriteria analysis. This involves the definition, hierarchical breakdown, and weighting of congruence criteria.

4.2.1.1. *Hierarchical breakdown and weighting of the congruence macrocriterion*

A pragmatic approach focuses attention on the immediate environment (proximity of content and near horizon). It is dialectically articulated around two points of view or criteria: utilitarianism and opportunism.

The concept of utility refers to the scope of the strategic intervention, that is, its appropriateness and functionality within the overall context and in particular for the actors. Timeliness is based on the admissibility and acceptability of this intervention by the global context and in particular by the actors of the politico-strategic context.

Each major dimension, thus considered as a macrocriterion, must in turn be broken down – or rather, interpreted – according to different components (technical-economic, legal-regulatory and politico-strategic). They result from the systemic representation of the context in the form of criteria, increasingly operative, to evaluate the congruence of the strategic intervention.

4.2.1.2. *Specification of interactions with the intervention context*

Pragmatism is based on the interactions of the strategic intervention with the different components of the intervention context. The interfaces involve the usefulness or appropriateness (or acceptability) of the intervention. They correspond, respectively, to the lower left line and the upper right column of the concrete intervention system (see Figure 4.3).

The first level of criteria focuses on two points of view:

– the usefulness of the strategic intervention for its overall context, reflecting the "impact of the intervention on the context" aspect;

– the appropriateness or acceptability of this intervention in its overall context, reflecting the "hold of the context over the intervention" aspect.

The second level of criteria seek to specify, respectively, the usefulness and the appropriateness according to the different subsystems against which they are evaluated. For example, in the technico-economic system, the usefulness of the action is broken down into its impact on regulation and its politico-strategic scope. At the same time, appropriateness is translated into technical/economic feasibility, legal/regulatory admissibility, and politico-strategic acceptability.

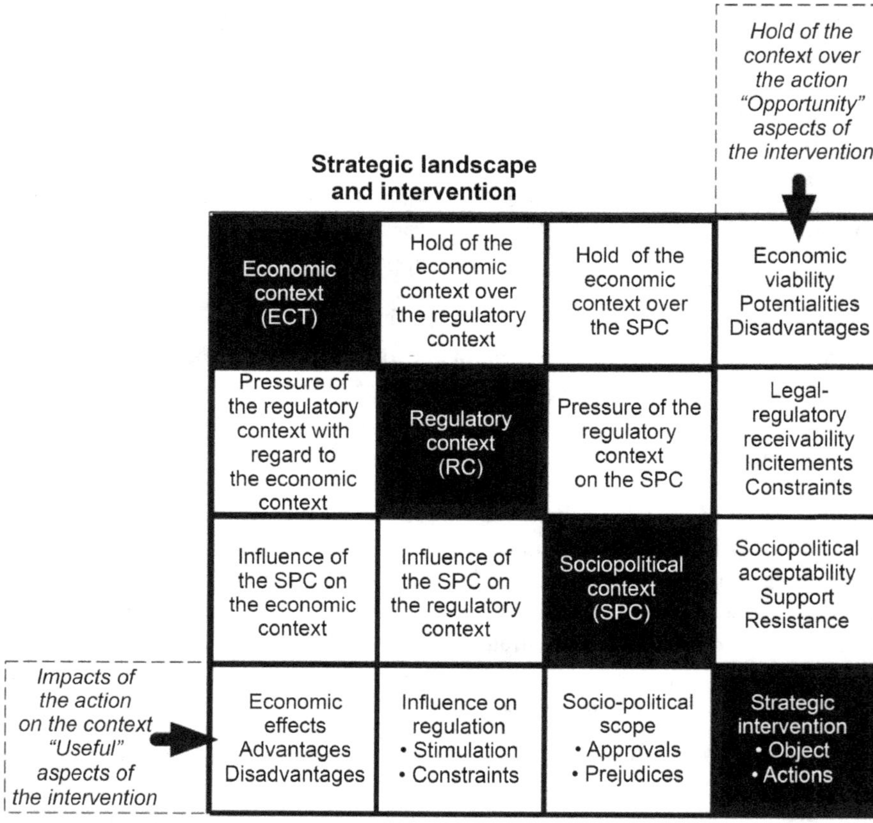

Figure 4.3. *Overall intervention system: grid for analyzing the congruence of the action-intervention with its "strategic landscape"*

Finally, a third level of criteria allow us to take into consideration the different components of the technical/economic subsystem, the primary actors of the politico-strategic context, etc.

For pedagogical purposes, it is important to focus, first of all, on the part of this framework relating to the sociopolitical context (SPC).

The pragmatism of strategic intervention is based on the congruence of the intervention with its context, which is to say, on its ability to integrate effectively into its environment. The environment consists of the actors involved and the way in which they shape the circumstances of the strategic action.

4.2.1.3. *Congruence of each option-intervention with the politico-strategic context*

This is a question of widening the angle of view by the representation of a meta-system – the "system of concrete action". This system places the intervention-action in its environment as well as in its broader context. The strategic intervention and its context are therefore the main diagonal blocks of this overarching framework. The interfaces between them reflect either the usefulness or the appropriateness (acceptability) of the envisaged action. These correspond respectively to the southwestern line and the northeastern column of this framework (Figure 4.3).

The first level of criteria is based on two dialectical points of view:

– the usefulness or the politico-strategic scope of the planned intervention;

– the appropriateness or acceptability of this intervention by this same context, that is, the politico-strategic pressure likely to be exerted by the actors on the intervention.

A finer breakdown (second level) of these criteria makes it possible, in particular, to consider the main actors of the politico-strategic context. Thus, the description of the interfaces of the envisaged action, with each of the actors appearing asymmetrically along two axes, is as follows:

– the usefulness and the politico-strategic scope of the intervention (southwestern line), which concerns the advantages (agreements) and disadvantages generated by the intervention;

– the political and strategic acceptability of the intervention, that is, the attitudes and reactions, or the pressure exerted by the actors with regard to the strategic intervention. These are expressed in terms of support or resistance.

The weighting method associated with this dimension is based on the relative importance given to the criteria at each level of the breakdown. However, it is necessary to note that a given actor does not necessarily have the same local weighting in the politico-strategic landscape depending on the perspective considered: the "contribution of the intervention" or its "acceptability by the actors". The politico-strategic pressure exerted on an intervention is all the more "readable" the more it is differentiated by the actor.

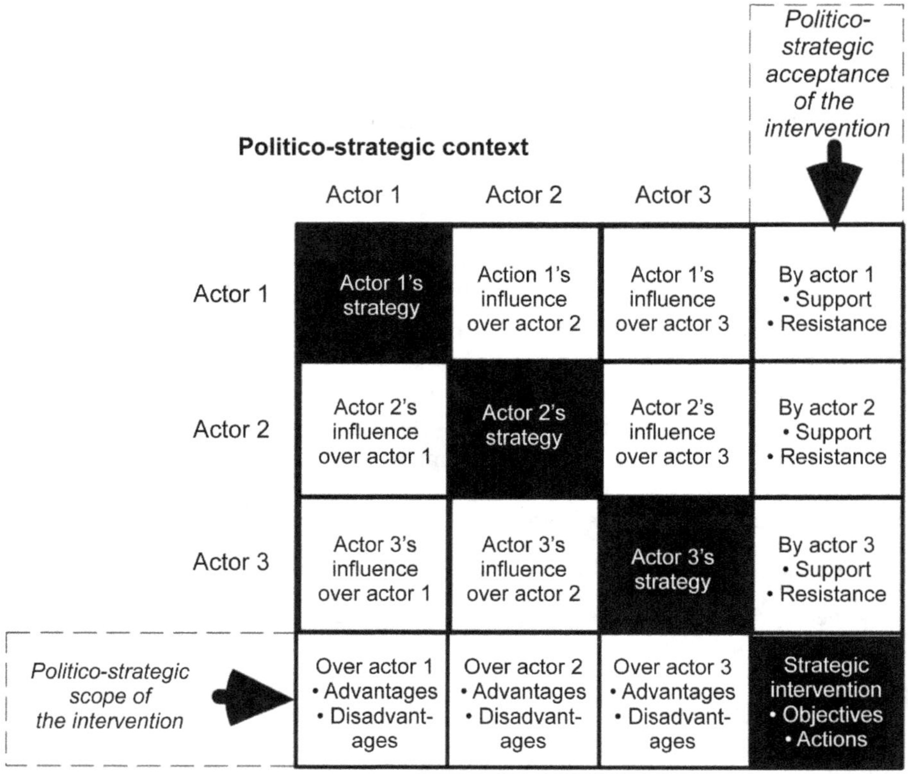

Figure 4.4. *Concrete system of action: analysis grid of the congruence of the intervention with its politico-strategic context*

The description of the interfaces of the planned intervention with each of the selected subsystems appears asymmetrically along two axes, depending on whether we consider the impact of the intervention on the context or the influence of the context on the intervention.

The influence of the intervention on its context, corresponding to the usefulness and convenience of the intervention, concerns, respectively:

– the economic effects, in the form of advantages or disadvantages;

– the influence on regulation, in terms of stimulating or hindering;

– the politico-strategic scope (amenities and prejudices), which has already been discussed.

The influence of the context (economic, regulatory, and politico-strategic) on the intervention, corresponding to the "appropriateness of the intervention", is assessed in terms of:

– the economic feasibility, through potentialities and/or vulnerabilities;

– the legal-regulatory admissibility, in terms of incentives and/or constraints;

– the politico-strategic acceptability (support and resistance) already mentioned.

More specifically, it is important to focus on the boxes in Figure 4.4 at the finest level of the grids to specify and assess the interfaces of the intervention with the subsystem (e.g. the politico-strategic context). These boxes, located at the crossroads between a component (for example, an actor) of the subsystem and a component of the intervention, include:

– a summary description (in the form of an item) of the highlights;

– a qualitative assessment of their favorable or unfavorable nature (support and resistance);

– a rating of their relative importance and their probable development.

This method for rating the strategic intervention makes it possible, at this stage, to identify:

– its *points of vulnerability*, based on the interface items presenting a major risk; in the particular case of congruence with the politico-strategic context, the primary inability to reduce prohibitive resistance (the "acceptability" aspect), as well as the harm caused by the intervention (the "appropriateness" aspect);

– its *assets*, based on the items that should be taken advantage of. These include, for example, the capacity to benefit from major support (the "acceptability" aspect), but also that of taking advantage of the advantages or approvals of the project for the stakeholders (the "appropriateness" aspect).

4.2.1.4. *Relative preferences associated with the options according to their sociopolitical pragmatism*

As a result, this method of assessing and filtering the intervention from the perspective of how congruent it is with its context makes it possible to revise it, in the sense of refining or improving it, and to determine the critical conditions and methods of implementation that cannot be avoided. It may allow for the alternative previously studied to be eliminated.

The comparative evaluation of individual solutions/actions is generally done in a table designed to compile the assessments according to each evaluation criterion, which constitutes the operational end of the decision tree. The table, attached to the branches of the tree, facilitates the scoring of alternative solutions and allows the calculation of a score for each of them. The score of an alternative solution or option is determined by the weighted sum of the scores, which it obtains for each of the operational criteria, where the weighting reflects the relative weight of the criteria in relation to the root of the tree.

Finally, this score is returned in the form of a numerical value, expressed as a percentage, and in graphical form. It can be used to interpret and analyze the sensitivity of the results through an appropriate software program. This result makes it possible to establish, through the comparative appreciation of the alternatives, the true scale of priorities in taking action.

Box 4.4. *Multicriteria assessment or comparative appraisal*

For each of the cells in the table, an *abstract* of the ratings is collected for the aggregate evaluation of the strategic intervention. In order to account for the asymmetric role played by these ratings, depending on whether they are favorable or unfavorable, the cumulative importance of favorable ratings (i.e. forms of support) and the absolute weakness of unfavorable ratings (i.e. resistances) is considered.

The weighting method associated with a given dimension of the context analysis is based on the relative importance given to the criteria at each level of the breakdown. Refining the breakdown of the subsystems has the potential to facilitate the assessment of the score for the intervention. Thus, the degree of refinement and precision of the constituent elements of the subsystems under consideration, in other words, the number of hierarchical levels retained for the analysis, may differ from one subsystem to another, due to the quality of the information available and the degree of interest of each subsystem with regard to the problem studied. In fact, specifying the criteria to make them operational allows for the method to be more transparent, and makes it possible for them to be assessed under a logic of choice. This is so that they may be better shared between the participants in the exercise, or also for their terms to be modified, while maintaining overall consistency.

4.2.2. *Assessing future validity as an indicator of future pluralism*

Scenarios are consistent combinations of plausible assumptions about the future, which are made with regard to major uncertainties. It is possible and desirable to perform a hierarchical breakdown of exploratory scenarios in the form of a tree on the joint assumptions about the future.

The scenarios are in fact constructed according to a two-step morphological analysis procedure. The first stage includes the identification of major uncertainties and their grouping into relevant thematic fields. The second stage concerns the elaboration of scenarios specific to each of the adopted themes. The second stage involves the consolidation of these in the form of global scenarios. At this stage, it is no longer a matter of combining the elementary hypotheses, but of combining the thematic scenarios, which are then considered as composite hypotheses. This aggregation procedure, in substance, consists of repeating the same method at a more global level.

4.2.2.1. *Transcription of scenarios into hypothesis-criteria trees*

On the basis of the scenario frameworks constructed in this way, each general environment scenario can be expressed as a cascade of key assumptions about the future, that is, a tree structure.

With reference to the example of the Crin club regarding the future of research in Europe, an illustration of this type of tree structure for the general environment scenario, "metamorphosis through innovation", is provided in Figure 4.5.

In this way, the breakdown into composite hypotheses (or thematic scenarios) constitutes the first level of this tree structure. The second level is then provided by the basic hypotheses, or underlying elementary configurations.

The future is multifaceted, and includes multiple possible general environment scenarios (GES). Each of these can form a branch of a broader tree representing the multiscenario future with all its virtualities. In this case, the distribution of the main GES branches expresses a *mutual exclusionary* relationship, while the cascade of assumptions constituting each of these main GES branches instead expresses a *conjunct relationship*. These GES essentially represent the spectrum of contrasted images in an uncertain future, to which the probabilities of them coming to pass will be assigned.

Such a representation of scenarios is useful, as it constitutes a series of hierarchical criteria which can assessed for how appropriate they are for future scenarios to assess their overall long-term validity, or to identify strategic actions. Each basic hypothesis, the foundation of a general environmental scenario, constitutes a specification or characteristic feature of that scenario. The contrasting of these actions with this significant aspect of the future reality can be qualified, and then evaluated, in terms of how appropriate they are for the scenario according to this hypothesis criterion.

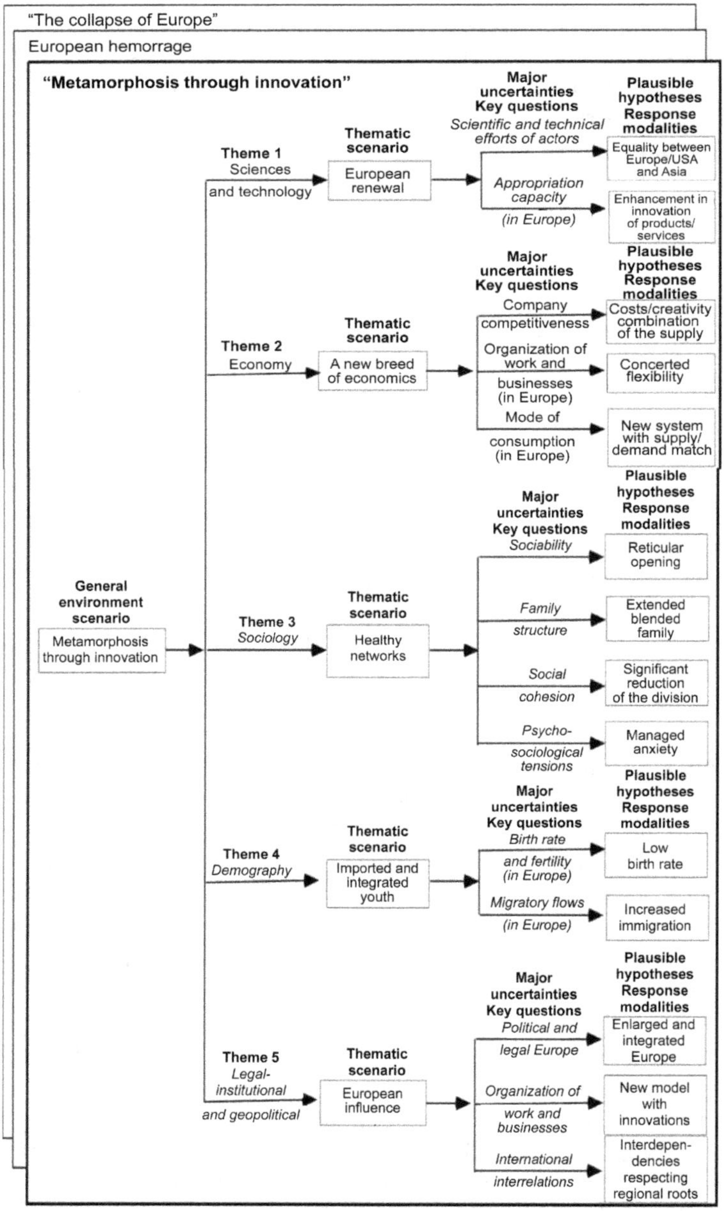

Figure 4.5. *Hierarchical breakdown of the "metamorphosis through innovation" scenario (according to the Club Crin Prospective scientifique et technologique)*

4.2.2.2. *Comparative validity of each option according to the exploratory scenarios*

The plurality of the profiles of hypotheses in the scenarios reveals the uncertainties to be found in the future. It is appropriate for the organization or company to anticipate through the adoption of a posture that takes this reality into account, thus the notion of the "pluralism of the planned intervention facing the future" or the virtuality of situational contexts.

From this perspective, the macrocriterion of pluralism should allow the assessment of the adequacy of the intervention to future environments. It is a question of putting it into perspective with regard to the different exploratory scenarios developed in the framework of a preliminary prospective study.

4.2.2.3. *Relative preferences given to the options according to their pluralism with regard to the future*

This dimension covers two aspects that interact dialectically, depending on whether the action shapes the future environment or whether it is subjected to its pressure: idealism and realism.

Idealism, which is the register of proactivity (while being only one of the many aspects of this concept), reflects the propensity of the envisaged intervention to inflect and shape the future environment by influencing the advent or the contours of one scenario or the other.

Realism is based on long-term positioning and the validity of the intervention in relation to the exploratory scenarios; it falls within the domain of pre-activity.

In accordance with the adopted systemic representation mode, the information qualifying the interfaces between the strategic intervention and the system/scenario under consideration is distributed as shown in the diagram of the concrete action system (Figure 4.4):

– in the upper right column, realism towards the future, indicating pre-activity;

– in the lower left column, idealism, relating to proactivity.

The tree structure of the subcriteria for evaluating the strategic intervention is the same for realism (pre-activity) and idealism (proactivity). However, both the nature and the determination of the weightings attached to these two dimensions are dissimilar.

Pre-activity focuses on anticipating the implications of each scenario in terms of the consequences (favorable or unfavorable) of the planned intervention, according to the different subsystems of the context (technical/economic, legal/regulatory, politico-strategic, etc.). It also involves assessing whether the planned intervention offers the project owner the opportunity to take advantage of the different scenarios likely to arise in the future.

From this perspective, where chance and necessity are predominant, the weighting to be applied must be based on the degree to which each of the scenarios could happen. In this respect, some methods, such as SMIC Prob-Expert, initially designed in 1974 (Godet 1997a, 1997b), allow calculating the degrees of likelihood of both general environment scenarios and thematic scenarios, based on "expert" estimates of the probabilities (simple and conditional) of their basic assumptions. However, these have proved to be inconclusive.

Proactivity, on the other hand, concerns the strategic capacity of a player to shape the environment in the long term, causing it to change in its favor. Thus, the entity planning the intervention may try to bring about one scenario or another, or attempt to influence its contours, in particular by causing changes in certain founding hypotheses. From this perspective, where chance and necessity may give way to will, the group of weightings must reflect the preference given to the different hypotheses underlying the considered scenario. The joint analysis method (Vernette 1987) appears to be the most suitable for this new weighting requirement.

The joint analysis method is used in marketing to select concepts or profiles of products and services. It consists of submitting to a group of experts – in this case the company's management team – a subset of profiles (or combinations) of hypotheses previously selected by an experimental design (in the statistical sense of the term).

The experts are therefore led to classify the different possible profiles of exploratory scenarios according to a qualitative scale of preferences reflecting their deliberate desire to see that some of them are carried out.

The method makes it possible to:

– estimate the preference attributed to all the hypothesis profiles, and in particular to those among the list of envisaged scenarios that could not be tested by the group;

– detect, through the highlighting of hypothesis profiles benefiting from a high overall level of preference, desirable scenarios that should then be considered;

– reveal, for each possible scenario, the partial preferences assigned to each of their founding hypotheses.

Box 4.5. *The joint analysis method (Vernette 1987)*

4.2.3. *Assessing the adequacy of the strategic reference as a guarantee of finalism*

The political and strategic frame of reference from which the "finality" of the strategic intervention is evaluated has two systemic components: on the one hand, the "vision", which represents its reason to exist, and which presides, through the finalities, over the global action in politico-strategic matters; and, on the other hand, the "fundamental abilities" which confer, in a sustainable way, a distinctive advantage to the organization over its partners, as well as a reasoned flexibility to face uncertainties.

4.2.3.1. *Relevance of the intervention regarding the major orientations*

The last major dimension concerns the macrocriterion of finalism. This dimension takes into account the relevance of the strategic intervention, both in relation to the vision of the organization in charge of the intervention, and in relation to its vocation or mission. This macrocriterion is underpinned by the dialectical composition of two criteria: voluntarism and conservatism.

Voluntarism is based on the adequacy of the strategic intervention to the vision. It reflects the qualification and validity of the strategic intervention considered as a response or solution to a singular problem in relation to the vision, of a structuring and permanent nature. It also concerns the sometimes unexpected consequences and repercussions of the strategic intervention with regard to the vision and/or the traditional missions of the organization.

Conservatism is observed through the conformity of the intervention to generic requirements that, beyond the vision as such, may be part of the mission, vocation, or culture of the organization. These requirements can sometimes take the form of consensual imperatives on aspects that are geopolitical, humanitarian, economic, social, etc.

These two criteria are complementary in the sense that, to be effective, it is necessary to know how to combine a desire to change with the pursuit of a certain continuity. Moreover, these criteria tend to be dialectically opposed. Indeed, by striving to respond to an important but singular problem, the organization in charge of the strategic intervention may be diverted from its original mission.

With regard to conservatism, the exercise consists of considering each of the objectives and components of each planned strategic intervention, and then qualitatively assessing the degree to which it conforms to the different facets of the vision. Thus, an interface matrix or cross-tabulation table between the intervention and the vision is filled in, making it possible to clarify the conformity of the various components of the strategic intervention with the main orientations of the vision.

This evaluation grid, depending on the preferences given to the different purposes, allows for a *multipurpose* evaluation of the conformity of each intervention to the vision or mission of the organization or institution.

With regard to voluntarism, it should be remembered that the facets of the vision only make sense if they are activated through strategic intervention. In other words, strategic intervention must contribute, in one way or another, to the evolution of the fundamental principles of the vision. The evaluation of the contribution of the intervention to the shaping of the vision is shown in an interface matrix that compares the intervention components (along the rows) against the facets of the vision (along the columns). This matrix identifies the directions and priorities of the vision influenced by the different objectives and components of the intervention. For each objective, the relative importance of each impact of this intervention is assessed on a qualitative scale, both for the priorities they are expected to satisfy and for all the terminal directions in the vision tree, including those not explicitly addressed.

The result of such comparisons offers, in particular, the possibility of obtaining feedback on both the strategic vision and the intervention. Indeed, it makes it possible to:

– proceed to the adaptation of the finalities of the vision with regard to their degree of activation by the interventions. For example, some goals, which would not be supported by any concrete action, can be adapted because they are considered too ambitious or revised downwards in terms of their weighting/preference;

– reorient certain interventions, or reconstruct them in an appropriate manner.

4.2.3.2. *A finalized and evolving design of the strategic intervention*

The different types of cross-comparison presented constitute feedback loops that make it possible to refine and flesh out the results obtained at each stage of the strategic engineering and politico-strategic decision-making process. These confrontation exercises also offer the opportunity to rebound on actions, strategic orientations and – to a lesser extent – on a vision and on more satisfactory scenarios, starting from initial solutions.

Thus, these cross-analyses make it more likely for the vision, orientations, and strategic interventions to emerge and become clear. The importance of such exercises goes beyond the sequential order of the normative determination of scenarios, visions, and strategic actions. From this point of view, it is better to have an initial foreshadowing of the vision prior to the scenarios, and then to contrast it against the latter, once sketched out. In a more pragmatic way, there is nothing to prevent starting from strategic interventions that would be confronted with a rudimentary vision. To a lesser extent, this vision will be revisited in the light of the concrete interventions and actions recommended.

4.2.3.3. *Relative preferences granted to the options according to their finalism*

Assessing how appropriate the intervention is to the core capabilities requires a consideration of their interactions. This involves filling in the confrontation matrices of the interventions or their components with the key abilities and distinctive know-how of the institution concerned.

This comparison, in terms of the intervention's capacity for insertion or "immersion" within its strategic landscape, is then considered from two eminently dialectical perspectives: the intervention's conformity to the fundamental capacity, and the intervention's contribution to the development or activation of this capacity.

By analogy with the scenarios, the fact of having used the morphological approach to formulate a concept of "fundamental capacity" makes it possible to rank it almost automatically in the form of a tree structure. The first level includes the key competencies selected by major domains. At the second level of this tree, the key competencies are broken down into distinctive know-how.

It is then possible to proceed to the cascading weighting of skills and know-how, keeping in mind that this weighting will be differentiated according to the assessment perspective (conformity or contribution to development).

4.3. Gauging the options according to their propensity to immerse themselves favorably in the strategic landscape: interpreting and appropriating the triad of macrocriteria

The three macrocriteria of pragmatism, pluralism towards the future and finalism form a triad.

These criteria are distinct, complementary, and sometimes contradictory. They are designed to allow the evaluation of the overall performance of strategic interventions, without obscuring their significant differences.

4.3.1. *A subtle mix of key priorities*

The triangle presented in Figure 4.6, the corners of which represent the poles of the triad, allows for a vision of proportion or the relative weights of the macrocriteria. The barycentric position of this composition of macrocriteria within the different zones defined around the center of gravity of the triangle allows the possible pre-eminence of a macrocriterion to be brought to light.

The process of strategic management is often inclined to favor finalism or pragmatism over pluralism facing the future, since the consideration of foresight is in most cases limited to the hypothesis of the realization of a given scenario. In the illustration presented using the triangle in Figure 4.6, the pre-eminence of pluralism tends to counter this tendency.

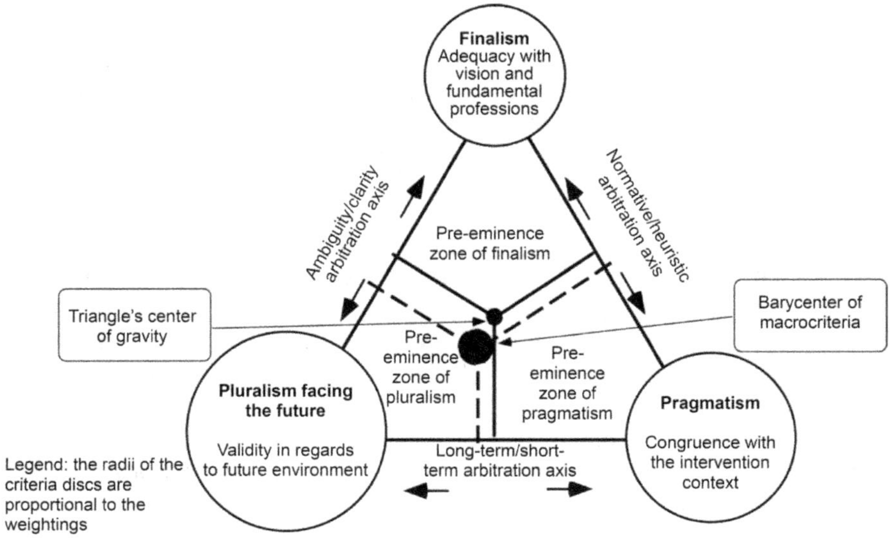

Figure 4.6. *The macrocriteria triangle: a fine balance of key priorities*

Moreover, the projections of this barycenter on the different sides of the triangle reveal the position of the planned intervention with regard to three arbitrations which, although implicit, remain no less fundamental:

– the arbitration between the normative and the heuristic or, according to Simon's (1991) thesis, between "substantive rationality" and "procedural rationality", is reflected in the position of the cursor of relative importance between finalism and pragmatism;

– the trade-off between the long term and the short term, which constitutes an indisputable front line between two conceptions of management, is expressed on the axis of pluralism in the face of the future/pragmatism;

– the trade-off between the ambiguity and uncertainty of the future and the clarity and uniqueness of the vision (creation of potential and resources), is revealed along the axis of pluralism in the face of the future/finalism.

The balance between these three macrocriteria, as well as its translation in the form of fundamental trade-offs, will evolve according to the nature of the interventions selected and the way they are tested.

4.3.2. *An evolving assessment of the intervention according to the macrocriteria*

Partial evaluations of the strategic intervention selected according to the three criteria of the triad (finalism, pluralism with regard to the future, and pragmatism) result in a distribution of scores, and in turn result in the aggregation of scores obtained according to more specific and operational criteria. The barycenter of the scores of the action thus makes it possible to position the action within the triangle of macrocriteria and to highlight possible imbalances in the evaluation of this action in relation to the relative importance given to the macrocriteria (weighting).

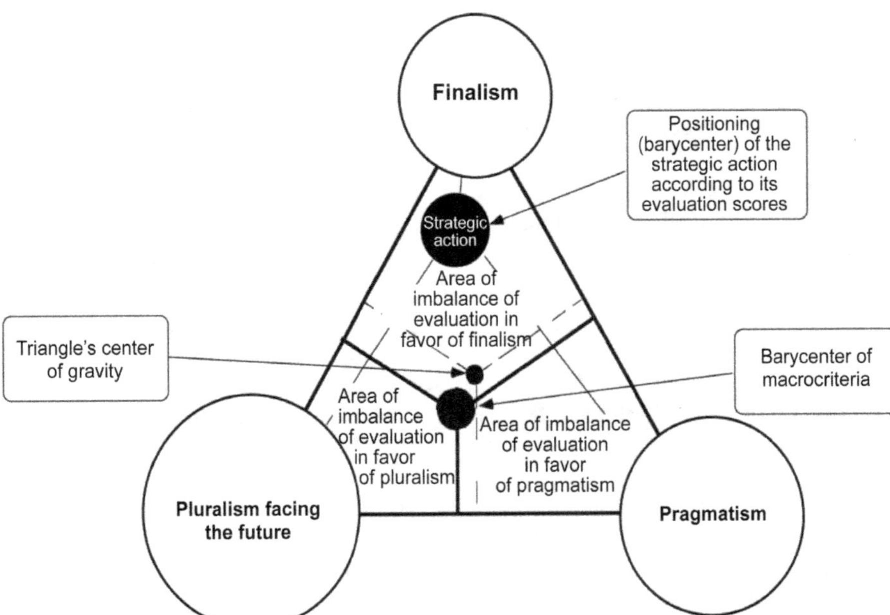

Figure 4.7. *Elements for interpreting the assessment according to the macrocriteria*

If we limit ourselves to an instantaneous evaluation of overall performance, it would be pointless to recommend strategic interventions where almost the entire scoring relates to the predominant macrocriterion, in this case, pluralism with regard to the future. However, such a logic cannot be sustained without the risk of seeing action become bogged down in turmoil due to possible refutation on the part of the actors (the lack of pragmatism and politico-strategic acceptability), or dispersed by a lack of clear orientations (lack of finalism, vision, or deliberate objectives). Admittedly, a distortion between the stated intentions through the weighting of the macrocriteria (the evaluation reference frame) and the positioning of the cloud of envisaged actions is necessary to "pull" the recommended intervention from a certain perspective. However, this discrepancy must not be so excessive as to be demobilizing. In this way, a mutual adjustment takes place between the weighting – and sometimes the content – of the criteria, and the deployment or evaluation of the action, according to each of them.

Strategists, when entering the field of action, cannot rely on an experimental method of trial and error. They need a repository of macrocriteria that will allow them to explore and influence the paths they have set.

4.3.3. *Using fundamentally circumstantial multicriteria assessment*

From a generic tree used as a reference frame for benchmarking, we can activate branches/criteria to describe a specific context during its operational use.

As shown in Box 4.6, this is a step away from the traditional framework of static trees to the more challenging framework of "dynamic" trees. This allows, almost automatically, for the recommendation of strategic interventions that can be revised, or even questioned, depending on certain critical changes in the context.

> Hierarchical analysis is the method used to model the reasoning applied to solve a given problem. It uses weighted multicriteria trees. The result is a hierarchical list of possible solutions, which can be analyzed and interpreted according to various representations using appropriate software tools.
>
> Depending on the invariant or changing nature of the situation studied, a static tree structure, or a dynamic tree structure will be used to describe and solve the problem. The latter of these, which is appropriate for changing situations, is the only one capable of representing the diversity of scenarios and of providing the representation (i.e. of experts) of a particular context likely to affect the final result in real time. Dynamic trees are multicriteria trees with conditional branches designed to support repetitive and real-time decision making based on recommendations (actions to be taken, measures to apply, risks to monitor, etc.).

Whereas a classic multicriteria static-type tree describes a context or a unique and unchanging reference point, a dynamic tree, due to its modular design, answers the question for which it was formulated in a differentiated way, based on the contexts to which it refers. Thus, any option for intervention at a given moment may be adopted in light of the conditions observed at that moment, and may well differ from the option advocated in other circumstances.

From a technical point of view, there is a virtual and exhaustive tree for all possible situations within which a selection is made. The gradual adaptation of the branches of the tree to the situation encountered is carried out interactively, that is, by means of an exchange between the user and the model, through the specification of appropriate parameters, upon stopping and switching points declared in the tree during its construction. Thus, a model can be rendered dynamic at any node of its tree structure.

By the choices they are led to adopt during the implementation of the model, users deploy a tree of criteria in real time through a "cascading" procedure, which is equivalent to selecting a dominant weighted sub-tree, through the pruning of possibilities that are not representative of the situation. In this way, the analyst/decision-maker progressively reconstructs a particular decision-making context. At the end of this process of questioning, the most relevant option or the one that fits the described situation is recommended. Additionally, the complete initial model remains available for the interactive description/selection of a new context and the prescription of solutions.

Through these conditional connections, expertise is transferred and relative weights are distributed – in proportion to the initial weightings – to the branches that are created. The solutions are then noted and weighted again in relation to the "new" tree.

This ultimately results simply from the instantiation-selection of a particular sub-tree within the generic virtual tree. The fact of specifying this sub-tree as the description of the situation progresses, via the "domino effect", towards gradually selecting scoring sheets, to reducing the set of scoring criteria retained, to modifying their initial weighting, which will ultimately modify the comparative evaluations of the various options or solutions. The natural result of this cascade of operations is the semi-automatic formulation of recommendations adapted to the situation and the context.

Box 4.6. *Dynamic trees*

5

Revealing Areas of Potential and Adopting a Decision-making Logic

It is difficult to make reasonable decisions without basing them on an adequate strategic posture, which should take into account the true nature and severity of the uncertainty prevailing within the current and future strategic landscape.

The context of the situation and the "theater of intervention" are littered with obstructions and shaped by constraints and rules that hinder or govern the capabilities of the company or organization to act. Thus, it is essential for strategic decision makers, before fixing any specific orientation and *a fortiori,* to have access to specific modes of action, a sort of "ultrasound scan" of the spaces within which they can reliably act with full freedom. The main objectives of this inventory consist of:

– leading strategic decision makers to accept these guidelines, within the framework of which they can carry out arbitrations;

– highlighting its performance potential;

– avoiding the many "optical illusions", some of which could prove fatal.

5.1. Modeling the decision problem and scrutinizing the spaces of freedom and opportunity

In order to reflect the preferences of decision-makers, in Chapter 4, we determined the relative weights given to the three macrocriteria (finalism, pluralism with regard to the future, and pragmatism). We then indicated how to initially establish a ranking of the strategic options on the basis of the weighted sum of their notes. There are many relatively sophisticated methods of aggregating ratings (Electre, Qualiflex, etc.) and preferences (Scharling 1985; Vincke 1989; Roy 1993).

For reasons of simplicity and priority, however, as opposed to such complex methods, decision-makers tend to instead prefer the more accessible method of ranking through "competition" using a multicriteria hierarchy. At this stage, we could stick to the choice of the strategic intervention judged to be the most satisfactory – and therefore preferable – according to all the criteria.

5.1.1. *Considering strategic initiatives in terms of composite options*

The proposed method is illustrated using the simplified case of a network services company in the telecommunications sector. In this case, essentially indicative in nature, the strategic tracks correspond to the reinforcement or development of structuring trades that have the possibility of constitutig the core competences of the company in the prospective horizon. This list of trades, presented in Box 5.1, is more concerned with "business lines" resulting from the combination of several areas of expertise, chosen from among the key areas by the company.

Possible deployment of seven fundamental or core businesses:

– operator of French infrastructures;

– export engineering;

– operator of cross-border infrastructures;

– product/service integrator;

– agency and intermediary;

– producer of value-added products/services;

– developer of patents.

Box 5.1. *Core businesses in a telecommunications company*

The ranking by "competition" would lead the solution with the highest overall score among the possible solutions being selected. Without denying the interest of such a classification, we should not be mistaken about the strategy issued. Beyond the choice of the option for the best candidate, this result is based on a very strong hypothesis, according to which these options are considered to be mutually or at least somewhat independent.

The problem of the decision may be limited to the exclusive choice of the most satisfactory candidate. Of course, it can also concern the "selection of first candidates according to the number of places offered in the competition". This may lead to the choice of a core of strategic interventions within a fixed budget envelope, for instance, or to the use of a limited platform of available human resources, or to various other constraints that may affect this choice.

It therefore seems essential to place oneself within the broader context of decision-making, in which the challenge is no longer to select the best competing strategic intervention(s), but to combine strategic options in the form of a weighted composition that best satisfies all the criteria. The real challenge for the decision-maker is to move from isolated and exclusive strategic interventions to true composite action programs. These would then result from the recomposition of all or part of these interventions.

However, such compositions cannot be envisaged without a certain coherence. To satisfy this last imperative, it is important to take into account, in the *form of constraints,* the delimitation of the "strategic field", the requirements and constraints inherent to the coupling of the options considered, as well as their implications for the company.

5.1.1.1. *Transcription of objectives and criteria*

Each strategic option or structuring action can be considered as a black box with inputs and outputs. This black box, for a combined level of operation, represents a column vector with the following elements:

– the unit output of the reference activity (*output*);

– the positive coefficients of use (*input*) of other activities or factors that may be limited (scarcity factors).

Each of them contributes to the achievement of the different objectives. The corresponding coefficients are called "unit contributions" of the action to the achievement of each of these objectives. The line vector of unit contributions for a given objective includes the performance evaluation scores of elemental or composite actions. This vector indicates, in substance, the orientation and the goal underlying the objective being pursued.

5.1.1.2. *Objectives-criteria matrix*

All the combined contributions of the actions to the different objectives correspond to the *matrix of objectives-criteria.* This matrix is the result of the comparative multicriteria assessment of the options considered (the table showing the scores of the different options according to all the criteria or objectives considered).

The unavoidable dimension for the design and evaluation of strategic options concerns the macrocriterion of pluralism with consideration for the future. This makes it possible to assess the suitability of candidate actions for future environments. The future, which is polymorphic, is in fact made up of a plurality of exploratory scenarios. It is therefore a question of evaluating the long-term validity of the interventions envisaged, with regard to these scenarios.

As an example, and for reasons of confidentiality, the summary description (Box 5.2) of the exploratory scenarios for the company concerned is limited to the broad outlines of their geopolitical and socioeconomic components alone, excluding the sectoral aspects.

– Mixed growth: sporadic and mixed recovery of growth against a backdrop of continued economic stagnation; a world structured in blocs, and a destabilized society.

– Mega-construction sites: proactively leveraging major pan-European projects and a strong international component benefiting from alliances and original financial arrangements.

– Social contract: a new model for society, emphasizing the qualitative content of growth, solidarity, increased cooperation, and individual fulfillment.

– Protectionism revisited: the stagnation of the world economy in an interventionist and sometimes nationalist and politically authoritarian context.

– Ultra-liberalism and globalization: rooting of growth in place, carried by new sectors in a liberal and globally regulated context.

Box 5.2. *Core businesses of a telecommunications company: five exploratory scenarios*

The objectives-criteria matrix offers the possibility of going beyond the "reductive" notion of a scenario that is central (or, at least, the most likely) and considering the range of contrasting scenarios.

Whatever recommendations emerge within the company in the form of strategic options, it is important to confront each of them with different scenarios.

Table 5.1 shows the most significant results from this comparison with regard to the first core business of an infrastructure operator at the national (or French) level.

This comparison is all the more important since, contrary to popular belief, the company's vision and missions do not stem from the results of the exploratory prospective. They may take the form of a table crossing strategic options in rows and operational criteria in columns. This table indicates the scope of each recommendation for action in terms of its future consequences, assuming that each of the scenarios and/or each of the founding hypotheses were to come true.

Scenarios for the future / **Strategic options (core businesses)**	**Growth**	**Mega-construction sites**	**Social contract**	**Protectionism revisited**	**Ultra-liberalism**
Operator of national infrastructures	Competition in the markets Decrease in productivity Limitation of solvency	Important gains, subject to obtaining large contracts	Reconquest of long-distance markets	Return to the monopoly against a backdrop of decline of international traffic	Fierce competition with spiraling price declines and collapsing margins
↓	↓	↓	↓	↓	↓
Associated performance scores (in hundreds of units)	1	3	1	2	– 1

Table 5.1. *Core businesses of a telecommunications company: qualification of the performance of the profession "operator of national infrastructures" in the context of each scenario*

The comparison of the business options with the different scenarios is shown in Table 5.2. Among other things, this table makes it possible to assess the relevance of alternative combinations of recommendations, according to the different scenarios considered, first in isolation and then as a whole. The idea is to be able to evolve from the notion of a choice of strategic initiatives, in the case of a particular scenario, to the notion of trade-offs, or even "bets", facing a plural future.

Such assessments can help to put the results of traditional economic and financial evaluations into perspective. Systematic use of the filter provided by this assessment grid can also lead to the reorganization, enrichment, and even reconsideration of certain recommendations or strategic orientations, with the goal of ensuring that they are valid over the long term.

Scenarios for the future / Strategic options (core businesses)	Growth	Mega-construction sites	Social contract	Protectionism revisited	Ultra-liberalism
Operator of national infrastructures	1	3	1	2	– 1
Export engineering	1	– 2	1	– 1	3
Operator of infrastructure outside the borders	2	– 1	1	– 2	3
Product/service integrator	2	– 1	1	2	3
Agency and intermediary	0	3	– 2	– 1	1
Producer of value-added products/services	0	2	– 1	1	– 1
Patent developer	2	2	1,2	1	2

Table 5.2. *Core businesses of a network enterprise: core business performance scores for each scenario (in hundreds of units)*

Even if a strategic initiative proves to be well-founded and relevant to the orientations of the vision (long-term positioning), and valid in light of the exploratory scenarios, it may be deemed unacceptable by the stakeholders and even discarded. This is why it is important to make the necessary adjustments to the strategic option, by making an effort to provide the best possible information, and if necessary, by lobbying the stakeholders of the initiative. This involves finding the best way to address the critical failure factors that could hinder the proper implementation of the recommended strategic option. The fact that a strategic initiative makes a significant contribution to the accomplishment of some of the company's objectives does not eliminate the risk of any side effects or negative effects that could prevent other objectives from being achieved. The effort devoted upstream to the prioritization of objectives cannot in itself guarantee the absence of structural contradictions between them. Likewise, the fact that a strategic recommendation is supported by a stakeholder, even an important one, does not exempt anyone from questioning the behavior of its possible detractors. This is why the exercise of "crosscutting" the consequences of any strategic option must focus on the various macro-criteria constituted by the objectives, scenarios, and actors, as well as on their respective subcriteria. These types of adjustments, revisions or reorienting of actions may potentially lead to an improvement in their intrinsic consistency. Making an initial judgment on the preliminary options for strategic interventions in the light of the criteria (objectives, scenarios, and actors) works toward ensuring their qualification.

To return to the calculation method, representing an option at a given level of execution consists of multiplying all its input/output coefficients by this same level (underlying assumption of factor proportionality or constant returns). Furthermore, the performance score of a composition of options on a given objective is equal to the sum of the unit contributions of these options multiplied by their respective levels of activity (additivity of contributions).

The various elementary actions, or black boxes, are then "nested" within the overall model. Thus, the whole system is an all-encompassing meta-black box, whose inputs and outputs respectively feed back into:

– the initial endowment (stock) of factors that are likely to be scarce;

– the expected overall results (objective to be optimized).

The simultaneous representation of the encompassing system and its constituent actions corresponds to the *constraint matrix of* the simulation-optimization model.

5.1.1.3. *Discernment of objective categories*

The objectives to be considered are of different types and can be classified into three categories or types, depending on their degree of "intentionality":

– the first type of objective includes the evaluation criteria, or *viewpoints*, against which the performance score is calculated. All the criteria resulting from the analysis of the strategic landscape are part of this category of objectives *a priori*;

– the second type of objective concerns the imperatives or *intentional requirements*. These are the criteria on which satisfaction thresholds are imposed. A threshold may, for example, result in a score that is considered disqualifying, depending on the criterion concerned;

– the third type of objective incorporates the *overarching goals*, or disproportionate ambitions, including the almost insatiable goals to be maximized, that is, those that one is striving hard to achieve. Such goals lead to an ever-increasing level of satisfaction on the part of the decision-maker and are accompanied by a "maximalist" type of decision-making behavior. These can be applied to the criterion goals in the first category. These goals may also be specific to the ambitions of decision-makers for the intervention, outside of any consideration of the strategic landscape.

Within a given category of objectives-criteria, composite objectives or criteria can be set. These are obtained by the weighted sum of the elemental criteria belonging to the same category. In this case, we face a multicriteria or multi-objective analysis. The weighting or preference system reflects the degrees of priority (or relative importance) given to the objectives or elementary criteria. These correspond to the probability of achievement (*influence/hold over*) or the level of "desirability" (*impact on)* of the scenarios.

5.1.1.4. *Formalization of the system of constraints*

The strategic landscape consists of three categories of constraints, classified in increasing order of complexity, corresponding to different types of concerns. Formalizing these constraints is an exercise that is as indispensable as it is delicate:

– the first category of constraints concerns imperatives or intentional requirements, corresponding to the ambitions for which a threshold of satisfaction is set. The coefficients of the intentional constraints are the ratings of the interventions. These are directly extracted from the matrices of interactions between the interventions and the strategic landscape. This category of constraints can also result from the imposition of a minimum weight or level required for certain components of the intervention (taking into account, for example, the inertia of the actions undertaken or "setbacks"). It goes hand in hand with the second category of objectives, already mentioned.

– the second category of constraints is based on the coefficients of use of resources or factors, as well as on the setting of thresholds concerning their availability (budget envelope, volumes of human resources that can be implemented by category, acceptable level of social feasibility of change, etc.);

– the third category of constraints, more complex to take into account, concerns the structural coupling relationships between the strategic options. The implementation of some of these requires that other actions be developed; some options may, for example, present synergies, incompatibilities, or cross-packaging.

In order for these to be made easier to formalize, the constraints in the third category can be deduced from the interaction matrices of the interventions. This automatic generation of coupling rules between options requires input from sources other than those of the projected actions. In order to do this, corrective coefficients for the constraints are introduced, depending on the "degree of self-sufficiency" of all the actions potentially to be included in the intervention. Thus, the consideration of partial self-sufficiency amounts to considering that the other actions are insufficient, as such, to reconstitute a given intervention. Thus, it is necessary to make the system more open in order to be able to provide itself with complementary factors likely to enter into the composition. This opening constitutes an additional space of freedom to recompose and choose solutions. Finally, in the proposed example case, it is possible to proceed to an almost intuitive formalization of all or part of the coupling constraints, as indicated in Box 5.3.

– *Floor or ceiling weight of certain core businesses*: for example, the relative weight of the "operator of French infrastructure" business is over 33%.

– *Limits on the use of resources or factors*: for example, there is a ceiling on the "overall conversion effort" (less than 3.30).

– *Compatibility and coupling of core business*:

- example no. 1: the sum of the relative weights of all the core businesses is equal to 100% (linked to the very definition of the core businesses),

- example no. 2: the relative weight of the profession "operator of offshore infrastructures" is greater than that of "export engineering",

- example no. 3: the sum of the relative weights of the professions of "operator of infrastructure outside borders" and "export engineering" must be within a range (floor 15% and ceiling 35%),

- example no. 4: the competitive pressure related to the use of certain trades must remain within an acceptable range.

Box 5.3. *Three types of simplified constraints*

Threshold constraints (floor/ceiling) related to the achievement of certain actions, the availability of certain resources (financial, human, etc.), or the imperative satisfaction of particular objectives are understandable by the decision-makers or actors involved. Taking constraints into account is made easier when they are tangible and close, that is, specific to the subsystem or category of factors concerned. On the other hand, coupling constraints (interdependencies, synergy, antagonisms, etc.) between options are more subtle and therefore more difficult to apprehend and formalize; hence the pedagogical interest of gradually introducing constraints, starting from the simplest and closest to the most subtle and remote.

5.1.1.5. *Formulation of the optimization problem*

Once the objectives have been modeled and the constraints formalized as indicated in the previous sections, the process of solving the problem of optimizing an objective consists of determining the nature of the basic options, in terms of selection and levels of shares or options, that have the best chance of accomplishing that objective. In this case, the objective is insatiable and must be pursued as long as the constraints allow.

When it comes to optimizing an insatiable goal, the decision-maker chooses from a range of goals that he or she considers to be paramount. This leads to the determination of the extreme level of the objective reached and the noting of the composite intervention associated with it. This must, by construction, be admissible, and therefore realistic, in the sense that it respects the system of constraints. The solution provides, among other things, the performance scores achieved on all the other criteria, as well as the actual level of use or generation of the related resources.

It then becomes possible to systematize the exploration, evaluation and choice of "composite interventions", that is, the compositions or bundles of strategic options discussed in Chapter 4. A composite intervention is characterized both by a particular selection of strategic options and by the distribution of the relative weights attached to them. This may provide an opportunity to present the current composition of interventions in terms of the actions being considered. In particular, it is intended to help decision-makers decide upon priorities among the almost infinite number of possible approaches, although some combinations are discarded for reasons of incompatibility.

5.1.2. *Revealing the scope of plural actions*

At this stage, it is important to conduct an "ultrasound" of the situation to be addressed or the problem to be solved, trying to clarify the contours without recommending a solution. Figures 5.1, 5.2 and 5.3, relating to performance windows, margins of maneuver and the setting of stress bounds, can serve as backgrounds or foundations for the triptych of decision-making dashboards.

The goal of this is to gain insight into and visualize the profile of the solutions or strategic options through these tables, under its different facets. This profile reveals the real spaces of freedom and incites one to reshape them judiciously:

– by delimiting and marking out performance windows, capacities for action, and areas of possibilities;

– by indicating the procedure to follow to favorably modify its contours.

5.1.2.1. *Visualization of windows of opportunity*

The method consists in revealing the actual performance ranges on the different goal criteria (objectives) as well as their true potential. These are the ranges of satisfaction that the company or organization can derive from each overarching ambition or goal, regardless of the composition of options that may be implemented. The management team can thus visualize and appreciate the practical value of pursuing each overarching goal, by taking into account its own concerns and ambitions.

This visualization is created using a "hierarchy of intervals", according to which certain goals reveal themselves, sometimes against all expectations, to be more favorable in terms of potentialities, or more discriminating due to the amplitude of their range.

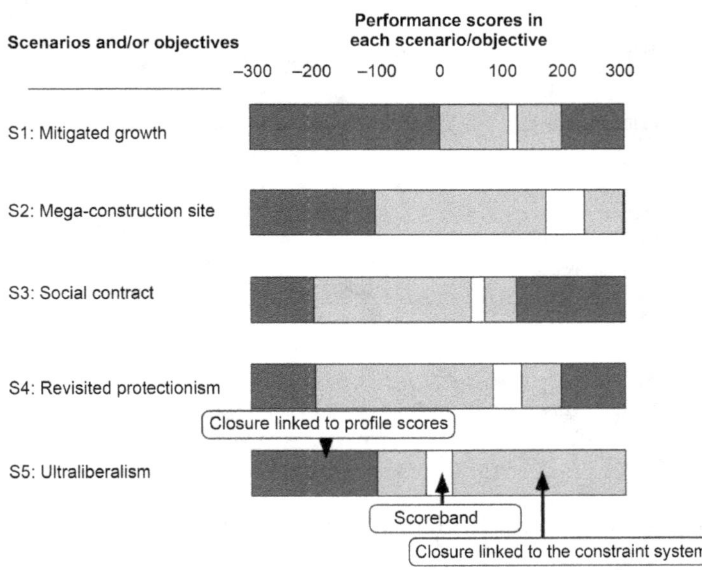

Figure 5.1. *Performance and implications by scenario and/or objectives: determination of performance ranges*

It is thus a way of internalizing constraints and clarifying their role by highlighting the effective performance they allow in terms of the company's various objectives. In this way, they also form "glass ceilings" that can hinder the achievement of ambitions; hence the interest that their anticipation has for the company.

Knowledge of these ranges makes it possible to calibrate each of the objectives by expressing it in the form of effective performance rates (or degrees). Evaluating the degrees of performance of any strategic composition in the form of percentages (without units) thus makes it possible to take into account plural ambitions, truly considering multiple criteria, even if the performance scores in absolute values for these different criteria are not expressed in the same unit.

5.1.2.2. *Delimitation of room for maneuver*

In addition, it is possible to delimit the effective maneuvering margins related to the composition of the shares or the deployment of the various options.

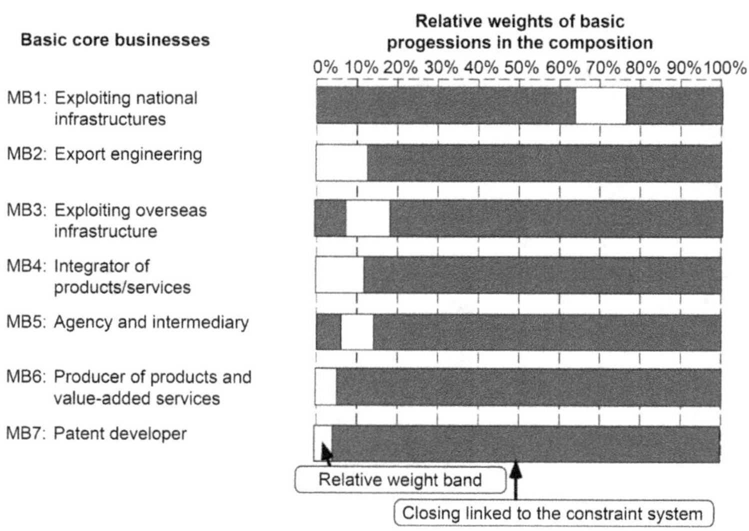

Figure 5.2. *Composition of the core business: maneuvering margins in determining the weightings of core businesses*

It is important to create a series of real and narrower windows that can be scanned through by the relative weight of the component in question, within the framework of compositions that respect the system of constraints.

These margins can be identified through the ranges of variation of the relative weightings of the possible elements in the composition. These weightings are supposed to be between 0% (excluded from the composition) and 100% (single and exclusive component).

The revelation of the floor and ceiling weightings is valuable information for the decision-maker. The decision-maker is given the maneuvering margins associated with each component of the planned intervention as well as the real possibilities for each of them to contribute to a realistic and coherent strategic profile.

This analysis of the floor and ceiling weights also makes it possible to measure the degree of effective activation of this component (in terms of additional weight) compared to that which would be imposed by the system of constraints. This visualization of the maneuvering room allows us to guard against optical illusions; for example, in cases where the relatively high weight of a component appears to reflect a particular effort on the part of the company, when in fact it is content to insidiously achieve the minimum required by the situation.

This approach offers the decision-maker the possibility of understanding the actual malleability of the constraints with a view to reshaping the field of possibilities.

5.1.2.3. *Marking of eligibility space due to constraints*

Each of the constraints is expressed mathematically by the existence of a limitation or bound (second member) to the validity of a value domain (first member). The decision-maker must be able to go beyond simply taking advantage of the straitjacket of constraints and requirements that are in force (reactivity) or anticipated (pre-activity), in order to be able to reshape his or her space of freedom (proactivity), to open up new horizons, while at the same time deploying his or her potential.

From this point of view, the proposed approach allows us to answer the question: how far could one reasonably claim to loosen a constraint, in order to increase its maneuvering margins?

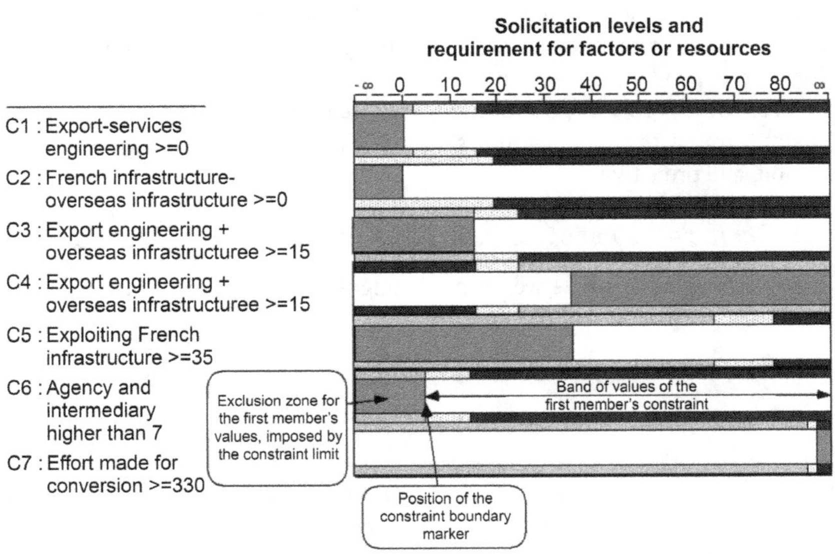

Figure 5.3. *Adaptation to requirements: areas of satisfaction and/or reshaping of constraints and requirements*

What is the limit to which the boundary marker for this constraint can be loosened – leaving all other constraints otherwise unchanged – beyond which the situation would not be resolved? The knowledge of the "incompatibility threshold" of the constraint system can in turn be completed and refined by the detection of another so-called "redundancy threshold". This threshold corresponds to the limit of loosening of the post, beyond which the constraint is satisfied, whatever the goals pursued or the solutions envisaged. Beyond this threshold, the constraint no longer has any effect. It is useless to try to release it, since the potential of the firm ceases to improve

5.2. Making wise use of uncertainty and ambiguity

Strategic engineering helps prioritize and select the most satisfying compositions of strategic options based on the categories of possible future attractiveness. When the different components of the problem of choice and their related treatments are sufficiently visible, this can help the strategist avoid unnecessary and even dangerous detours.

5.2.1. *Adopting a strategic posture according to the level of uncertainty*[1]

The methods for determining scenarios, and consequently the strategic postures (long-term vision or positioning) as well as the premises for intervention, can differ significantly depending on the type of environment in which the intervention is to take place. Moreover, the postures proposed echo four decision-related registers, articulated around the main attitudes towards the future: reactivity, pre-activity, prevention, and proactivity.

5.2.1.1. *Characterization of the main types of environments*

From this perspective, we can distinguish four main types of future environments:

1 In October 2001, this section was developed by the author as part of a conference at the AFPLANE (*Association française de stratégie et de développement d'entreprise*, French Association of Strategy and Business Development) on the subject of "Articulation prospective, stratégie et veille : vers une nouvelle approche du risque stratégique" ("Prospective interconnections, strategies, and oversight: toward a new approach to strategic risk"), as well as being a contribution to a book by multiple authors entitled *Comprendre et gérer les risques*, published by Éditions d'Organisation in 2002.

– *Continuous environments* are characterized by the pre-eminence of trends. Prospective exploration is limited to focusing on the duly identified trend scenario, the so-called "framing" scenario; in this type of environment, uncertainty is considered as residual.

– *Discontinuous environments* incorporate many possible ruptures. However, these are only projections, owing to the lack of precision on the circumstances or the date of their advent, due in particular to their complex causalities. They can, for example, result from the reaching of a limit in the regular evolution of certain variables, such as the progression of production volumes with saturation of the industrial tool leading to decreasing yields. In this type of environment, the future cannot be summed up in a single scenario (even if it is a trend), but rather in a range of contrasting and probable scenarios to which we can assign probabilities of occurrence.

– *Unweightable environments* conceal an almost infinite number of possible futures. However, these can be considered as being marked out by a few relatively identifiable – but impossible to balance – extreme scenarios. In this type of environment, future outcomes result from anarchic "mixes" of extreme scenarios, the weighting of which is unknown.

– *Indistinguishable environments* are characterized by the fact that they are subject to disconcerting upheavals, due among other things to the moving field and the versatile nature of possible futures. These types of environments consist of near total uncertainty.

5.2.1.2. *Recommendation of adapted decision-making postures or logic*

Each of these environments requires prospective exploration methods, strategic postures and premises of the intervention that are adapted, which are important to review. These postures, as well as the corresponding attitudes, must take into account the type of environment and the levels of uncertainty that is faced, namely: *continuous* (with trends dominating), *discontinuous* (with possible and anticipated ruptures), *unweightable* (an anarchic mix of scenarios), and *indistinguishable* (disconcerting upheavals) environments.

In the case of a continuous environment, the strategic posture or decision-making logic that should be adopted is to deliberately rely on the trend scenario.

But this posture may turn out to be an even more daring, given that there is a chance this trend scenario will not come to fruition. The vision (long-term positioning) then consists of internalizing this scenario, of trying to bring it about, while trying to redraw or clarify its outlines. As for strategic orientations and interventions, they should focus on the application of this framework in terms of objectives, action plans, or support measures.

In the case of a discontinuous environment, the most appropriate strategic posture is to opt for a robust compromise on the future, taking into account the many different scenarios and the possible likelihoods that they will occur. As for the strategy, it is the result of a judicious mix of strategic actions, some of which constitute responses adapted to this or that scenario, while others may come from various imperatives (deliberate will, commitment already made, actions underway, etc.). These actions will be accompanied by optional measures depending on the scenario for future outlooks.

In the case of an unweightable environment, where fighting against adversity is given priority, the lack of reassuring information prompts us to protect ourselves from any unpleasant surprises in the future. The strategic posture consists of facing the possibility that such surprises may in fact come to pass, by means of a preventive locking in the face of the contingencies. This translates into the adoption of a cautious common core of options (in the sense of game theory), such as to guarantee an acceptable performance floor, whatever the future may bring.

Objects of the strategic perspective / Environment type	Exploratory Scenarios	Strategic positions	Strategic actions
1) Continuous environment **Preeminence of trends**	Predominance of a "trend" scenario	Paris takes a bold approach on the trend scenario	Cluster of actions provided in the framework of the trend scenario
2) Discontinuous environment **Possible and anticipated breakdowns**	Some contrasting and plausible scenarios	Robust compromise of different scenarios	Multiscenario mixed future strategy, supplemented by optional shares
3) Unweightable environment **Anarchic scenario remixing**	An almost infinite number of possible futures marked out by extreme scenarios	Preventive locking against "unweightables"	"Prudent" base of shares guaranteeing an acceptable floor for the performance whatever the future
4) Indistinguishable environment **Disconcerting upheavals**	Moving field and versatile nature of possibilities	Learning skills and rules of conduct to best manage surprises	New key competencies with hyper-responsive interventions along the way

Table 5.3. *Interconnection between exploratory scenarios, postures, and strategic actions according to the type of future environment*

In the case of an indistinguishable environment, with near complete ambiguity to the point of being easily confused, the strategic posture consists of placing oneself in an overhanging position (a "metaposition"). This corresponds to a higher order of abstraction that reduces complexity by relying on forms of stasis. In this type of environment, which is often transitory as new paradigms emerge, the factors of uncertainty change over time and the very nature of the scenarios is called into question. We must therefore focus on the development of learning capacities that encourage the adoption of rules of conduct that allow us to manage the surprises in the future in the best possible way. At the same time, it is important to acquire fundamental skills (rewarding and distinctive, sustainable, malleability generating). Along the way, these will be accompanied by hyper-reactive interventions.

5.2.2. *Exercising circumspection in an almost inextricable situation*

It is clear that most of the interventions are based on planning (known as "strategic" planning). However, according to our classification proposal, the underlying strategic posture unquestionably refers to continuous environments. In this posture, prospective exploration is essentially limited to relying on a long-term framework, from which coherent action plans can be developed.

5.2.2.1. *A relatively frequent and sometimes fatal misdirection*

On the other hand, discontinuous environments – and ones that are *a fortiori* unweightable or even indistinguishable – are less familiar. However, they increasingly represent the reality, due in particular to the phenomenon of globalization, the speed of change (technological, geopolitical, regulatory, economic, etc.), the rise in complexity, and the multiplicity of interdependencies. As the environment becomes surreptitiously discontinuous, or even totally uncertain, we persist in adopting prospective exploration approaches, decision-making logic, and modes of intervention that are mainly based on continuous environments. This optical illusion, given its adventurous and sometimes chronic nature, will continue to compromise and further mortgage the future of many companies and organizations.

5.2.2.2. *Temporal evolution of the degree of uncertainty and differentiation according to the object of study*

Although considered exclusive and unique in the context of a given subject of reflection or study, the different types of environment inherent to the degree of uncertainty may change over time. Thus, it is in the interest of the company or organization concerned to revise – if necessary – the strategic posture or decision-making logic initially adopted, to counteract this evolution. Furthermore, these types of environment, defined according to the degree of uncertainty that prevails, are obviously differentiated according to the countries, sectors, areas of activity, or issues concerned.

5.2.2.3. *Conjunction of environmental types and building a hybrid posture*

Within the framework of a particular strategic reflection, if the type of environment is observed in a vigilant, rigorous, and permanent manner according to the degree of uncertainty in the environment, it should manifest itself at all times in a sufficiently divided way.

Contrary to the conclusions drawn from the previous sections, the environment does not always present itself unambiguously. Indeed, it sometimes can be found to take many forms, and is characterized, in this case by the relative conjunction or synchronization of the different types proposed. This fifth and last type of environment, which can be described as composite, is then more confusing and difficult to grasp. It can therefore be considered to consist of different layers of major uncertainty factors, respectively associated with the four types of environment defined above (continuous, discontinuous, unweightable and indistinguishable). Faced with this new *state* of exacerbated *uncertainty*, the approach consists in identifying, for each of these strata, an appropriate fragment of strategic postures and actions, inspired by the proposed typology. In this way, a final hybrid type of strategic posture may be constructed. This posture toward postures (of a metaorder) would then, by construction itself, be stratified through the stacking of fragmentary postures associated with increasing degrees of uncertainty.

In conclusion, the foresight and strategic anticipation approach cannot be applied to an environment that is too indistinguishable or fundamentally chaotic, without experiencing difficulties. This is why the strategic or decision-making paradigm that should be adopted in order to be truly cautious is all the more subtle and complex, as the environment appears more perilous for the organization.

6

Recomposing Solutions and Making the Best Decisions

To be satisfied with the intermediate results of the multicriteria comparative assessment in order to adopt a particular option would be no better than "stopping in the middle of the ford" – truncating or cutting short the normative process of maturing choice.

A rigorous consideration of the constraints and rules in force in the context, inherent to the reality and complexity of the situation, often leads leaders/strategists to more informed decisions in their arbitration. In most cases, these decision-makers are in fact brought to reject the exclusive choice of one of the options suggested by the multi-criteria analysis, instead favoring a more realistic intervention/compromise resulting from a judicious balance between such options.

From this point of view, the purpose of this chapter is to propose a fruitful combination between simulation and optimization, by presenting an instrumented method for supporting decisions under constraints, complexity, and uncertainty.

The instrument presented in this framework highlights and deepens the classical mechanisms: from optimization, using simulation as an engine for configuring the problem and its solutions, to post-optimization. However, it gives a more discrete role to optimization, as a vector of simulation efficiency, particularly for exploring the real areas to maneuver. The virtue of this second function is that it encourages decision-makers to take on constraints and requirements that act as a straightjacket. This appropriation also makes it possible to reveal the stimulating challenges that these strategists can address, by trying to judiciously bend these requirements.

6.1. Reconciling requirements and risks by combining simulation and optimization

The software tool developed for this purpose, Simoptis, is summarized in Box 6.1. It is structured around an original concept based on a functional conjugation between simulation and optimization in uncertainty and complexity.

> Simoptis is a simulation-optimization tool for strategic solutions. It enables the recomposition of the best judicious options, making it possible to decide wisely in situations of complexity and uncertainty and to deploy coherent and efficient actions.
>
> Simoptis is based first on a comparative evaluation of strategic options on the basis of multiple objectives, considered as multiple criteria. In order to do so, approaches allowing the generation of criteria trees and their cascading weighting are used. These take into account the subjective and qualitative nature of the decision-maker's preferences within the framework of a classic scoring procedure.
>
> However, this prior "competition candidate" comparison, in order to be fully satisfactory in terms of choice, is based on the prohibitive assumption that these options are considered to be exclusive of each other and relatively independent of the constraints inherent in their environment. In order to overcome this hypothesis and better fit the concrete situation, Simoptis allows the strategic options to be recombined and selected, taking into account not only the wishes of the decision-makers, but also the multiple requirements and constraints they face. These realistic recompositions are carried out on the basis of a range of decision-making logics, representing different strategic positions (reactive, bold, robust, prudent, proactive).
>
> Simoptis also makes it possible to reveal the real room for maneuver and to indicate the sources of potential or increased capacity for action, while indicating the related strategic risks and vulnerabilities.
>
> Among the many other fields where Simoptis can be used are the prevention and application of the precautionary principle, incentive policies and compensation measures, management of product and service portfolios, governance of large investment, or R&D programs.

Box 6.1. *Simoptis: simulation–optimization of strategic interventions*

6.1.1. *Conducting a cascading arbitration and selection process*

The decision support process is threefold, as it enables navigating through a framework of choices taking the form of "Russian dolls":

– a trade-off between different strategic or logical postures of choice that have already been presented, the respective fields of relevance of which correspond to the degree of uncertainty attached to the future environment;

– the adoption of strategic structuring postures;

– the establishment of basic selection premises or parameters (e.g. among a range of objectives) to enable the appropriate development of "innovative cocktails of intervention".

By exploring and gauging different cascades of register–posture–premise–intervention choices, decision-makers can make the best choice in a composite intervention.

The operational objective of the software is to compose adapted solutions in order to make efficient decisions. From this perspective, Simoptis offers the possibility of exploring and evaluating a multitude of composite solutions (or compositions of interventions) that meet the requirements and constraints of the situation.

Simoptis also helps to prioritize and select the most satisfactory compositions of strategic interventions using a palette of "decision logics". These logics provide an operational content to the three categories of possible attitudes towards the future: reactivity, preactivity, and proactivity. As explained in Chapter 5, the choice of these attitudes is influenced by the type of environment that the company or organization faces.

Within each logic, Simoptis provides a particularly pertinent grid for the analysis and interpretation of results.

6.1.2. *Adjusting inconsistent interventions by adapting them to constraint*

The first type of optimization is a normative approach in which a guideline is set, or even imposed, in terms of an ambition or overriding objective. However, this approach can prove unrealistic and give rise to specifying or revisiting a double result in terms of the target reached and the admissible composition of activities.

The second type of optimization, less self-absorbed and less ambitious, attempts to address the problem raised by the relative incoherence of certain compositions of pre-existing interventions.

It sometimes happens that the progression of the constraints calls into question the validity of an intervention. Moreover, many of the recommended composite interventions may *ultimately* prove to be unrealistic. In the case of strategic compositions that are not coherent – that do not respect the constraints – the simulation-optimization tool helps to make the recommendation coherent by determining a counter-proposal that is better suited to the constraints, while respecting the implicit concerns of the decision-maker as much as possible. This involves providing guidance on which criteria and to what extent decision-makers will have to "let go" in order to become realistic by complying with all the constraints.

These reactive adjustments to constraints can concern:

– a realistic counter-proposal based on a recommended intervention;

– bringing a previous position or an ongoing intervention back into conformity.

During the consistency process, decision-makers may indeed be confronted with a pre-established composition of interventions. This composite proposal for action may arise from their own vision of the intervention or come from the recommendations made by other actors, or even from an injunction from their hierarchy. But regardless of how it originates, this proposal can only be considered relevant if it is admissible – in other words, if it respects all the constraints that are imposed, independently of any particular desires of the actors. If this unavoidable condition is not respected, the decision-makers will no longer seek to optimize an objective that is specific to themselves, but will merely be content in respect the characteristics of the recommended intervention as best they can. To do this, they will try to adjust the intervention, trying to make it coherent and realistic, while trying not to distort it.

From a conceptual point of view, their approach consists in determining, from among the compositions of eligible interventions, those that deviate the least from the initial proposition, which amounts to minimizing the distance[1] between this initial proposal and the one they are seeking to recompose. In this case, it is a particular optimization function, in which the objective to be optimized is not a primary goal, but rather a degree of reconciliation between two composite interventions. In addition, decision-makers will be inclined to modulate their adjustment according to the wishes of the internal structure's hierarchy or the degree of adherence to the different levels of the actions recommended in the initial composition. From this point of view, they will have to take into account the degrees of firmness that should be assigned to the different levels of activity, in order to concoct a counter-proposal that is as close as possible.

Thus, the decision-maker integrates the initial recommendation and strives to spare susceptibilities by relativizing the extent of their adjustment. In the absence of primary goals or other deliberate objectives, the decision-makers refer to a "default",

1 Relative to a given standard, such as the sum – which may be weighted – of absolute values.

second-order objective, intended to ensure that the initial intervention proposal is brought back into coherence – through an adaptation to the constraints. However, the pursuit by decision-makers/strategists of this particular optimization logic does not deprive them, despite all of this, of any room to maneuver. In fact, through the interplay of the *coefficients of firmness*, they express a preference for a given recommendation of activities; which corresponds to a rule of conduct in the adjustment, or to a means to address the problem of the non-admissibility of their intention to act. This ultimately leads to the realistic fixing of a globally satisfactory counter-proposal.

6.1.3. *Developing a robust trade-off in a reasonably uncertain future*

An initial logic, known as the "pre-active realization of deliberate ambitions", makes it possible to optimize the relevance of the strategic position in relation to the objectives and/or the adequacy of the company's objectives to the scenarios. This logic is more precisely comprehensible in terms of:

– a bold gamble on a particular ambition (for example, for a desirable scenario, considered – perhaps wrongly – to be almost certain);

– a robust trade-off against several ambitions (e.g. for a probabilized set of scenarios).

6.1.3.1. *Pre-active realization of deliberate ambitions*

According to this logic, solving the optimization problem of a given objective consists in determining the composition of elementary actions, in terms of their selection and relative levels, in order to best accomplish this objective, which in this case is considered as the primary goal.

The so-called "optimal" solution is a particular combination of actions that must be realistic – that is, that respects the system of constraints – while providing a maximum degree of attainment of the objective pursued or the achievement of the performance potential.

The solution recommended by the Simoptis tool specifies, among other things:

– the relative levels or weights of the actions that make up the intervention;

– the input/output flows between the different activities, as well as the consumption of resources required for the composite intervention;

– the ultimate level of satisfaction achieved for the primary objective or goal;

– the solution's performance scores on the various other critical objectives.

6.1.3.2. *The multidimensional space of solutions*

A composition of interventions can be represented in a space that contains as many dimensions as the number of elementary actions or interventions that have been projected. The composite intervention is represented by a point in this space, whose coordinates on the different axes correspond, respectively, to the level of activation or to the relative weight of the action concerned in the composition. In order to facilitate a decision-making interpretation, this same composition will be represented by the simultaneous plotting of coherent morphological profiles.

Constraints may sometimes be mutually incompatible. In this case, there are no solutions. Therefore, it is useless to try to find them in the register of usual methods. In this case, any solution could only be the result of an effort to go beyond one's own limits. It would also be necessary to become aware of the problematic situation to be resolved in time, and of the severity of the decision-making challenge to be met.

In some cases, the space for solutions may be limited to only one possibility. This single solution is independent of the optimized objective and leaves no room for maneuver to decision-makers.

When the space of the solutions is consistent, it is multidimensional. As an analogy, this space can be compared to a polyhedral form, whose faces would represent the constraints. In this case, the optimal solutions are affected by "attractors" located at the tops of the polyhedron. This is why the most common procedure for searching for optimal solutions, called the simplex method[2], consists of successively exploring the different vertices, while increasing the degree of accomplishment of the primary objective or goal that one is trying to optimize.

At the optimum level, when the resulting solution best satisfies the objective being pursued, only the constraints that hinder the achievement of the overriding goal should be counted. Other constraints are considered inoperative, since they do not affect the optimal composition. Thus, the degree of saturation of the constraints is contingent on the optimized goals.

As we can intuitively imagine, the addition of non-redundant constraints reduces the space for the solutions and, consequently, the company's room for maneuver. Generally speaking, in a pictorial way, this addition amounts to shrinking the polyhedron of solutions, while increasing the number of its facets, and therefore its edges and vertices. Moreover, this type of addition tends to complicate the mix of solutions.

---

2 The simplex method is a technique and an algorithm for solving linear optimization problems faster.

A multi-objective optimization of strategic options without considering constraints provides a solution by performing only the most favorable (or best rated) option in a suitable multicriteria analysis. However, as additional constraints are introduced, pure (or 100% accomplished) solutions prove to be unrealistic and will need to be blended with other options. However, these initial options can be seen as the starting point for a process of convergence towards a realistic solution. These are in fact based on a spontaneous, watered-down, and somewhat utopian understanding of the situation, due to their ignorance of the real restrictions, or the failure to take into account the constraints and requirements at work in the strategic landscape.

Mixing solutions tends to moderate the achievement of goals/ambitions. In fact, the need to combine the initial options to find a realistic solution results in a reduction in the range of performance of the solution on the optimized objective.

6.1.3.3. *The reference price as the degree of scarcity of the resource*

The simulation-optimization tool indicates the pressure exerted by the various saturated constraints which hinder the achievement of the primary goal. These pressures more precisely translate the "failures to perform" or performance defects of the goal, attributable to the constraints. In economic jargon, this notion corresponds to the "marginal benefit", in other words, the value that the optimized goal could increase if this constraint were relaxed, such as by increasing the availability of the resource being considered by one unit. This notion is also equivalent to the maximum value (measured in terms of the objective) that one could reasonably be expected to concede to obtain an additional unit of resource; hence the name "shadow price" or "fair price" (in terms of the objective equivalent) associated with the acquisition of the resource affected by the constraint.

This price also reflects the degree of scarcity associated with the resource under consideration, in the pursuit of the broader goal. In less technical terms, it indicates how sharply the availability constraint impedes the achievement of the optimized goal. This price reflects more precisely the extent to which an additional unit of the resource in question could improve the ultimate level of this insatiable goal, in order to increase the performance potential of the company. An essential feature of the reference price is that it incorporates most of the interdependencies (direct or indirect) of the shares and resources solicited, as well as the limitations that govern them. Finally, the ranking of resources according to their reference or scarcity prices reveals the implicit hierarchy of constraints, depending on the pressure exerted on the composition of the optimal solution.

6.1.3.4. *The marginal cost as a ticket to enter into the action*

Each action considered in isolation has a microeconomic benefit. This represents the combined contribution of the action to the achievement of the optimized objective. This same action or intervention has a corresponding cost, known as the “marginal” cost, related to the use of the various factors, and limited or even scare resources.

This marginal cost can be evaluated by simply valuing the effective quantity (unit coefficients) of the factors used at their shadow prices. The mesoeconomic benefit is then obtained, which is not limited to the restrictive consideration of the action in question. This benefit integrates both the interdependencies of the actions and the scarcity of their factors and resources, via the difference between:

– the microeconomic benefit or the unit contribution of this action to the objective;

– the marginal cost of the factors that this action requires, which is valued at the reference or shadow price.

At the optimum point, that is, when a composition of actions has been determined that allows the ultimate level of the overarching goal to be reached, we can also provide the so-called “marginal substitution coefficients” between actions. These are the relative volumes of activity that need to be added or removed to reconstitute a given action at the unit level.

The marginal cost of this same action can then be calculated in another way: by valuing the activities added or removed by their respective microeconomic benefits. At the optimum point, the mesoeconomic marginal benefit of the retained actions is zero, while that of the eliminated actions is negative. The related cost represents a kind of “entry ticket”; that is, the fair compensation of the decrease in the value of the objective level that results from the forced introduction of an action not selected at the optimum, therefore considered undesirable. The evaluation, if one is made, of this cost for any new action makes it possible, among other things, to deem how appropriate it is, without it having been explicitly integrated into the model.

6.1.4. *Guarding cautiously against contingencies and any eventuality*

To facilitate the understanding of this particularly complex issue, we will limit ourselves to the framework of general environmental scenarios (GES).

The goal here is to identify cautious intervention strategies (in the sense of game theory) that can guarantee a minimum level of results, regardless of the possible scenarios that may arise. By choosing a conservative composition, the decision-maker can guard against possible future setbacks.

Determining the composite intervention that is consistent, that respects the constraints, is likely to provide a guaranteed minimum satisfaction (GMiS).

In this framework, the average satisfaction corresponds to the mathematical expectation of the scores achieved on each of the scenarios, weighted by how likely each one is to occur. To determine this cautious or prudent intervention, the calculation procedure consists in scanning an infinite number of situations concerning both interventions and scenarios.

6.1.4.1. *Prudent strategy in terms of satisfaction*

The goal is to choose a particular composition of interventions, known as a "reference" composition, and to test the satisfaction it would provide to the company for each of the possible combinations of scenarios. The idea is to identify the scenario combination corresponding to the minimum score (*Min*) for this reference composition. The same exercise is then repeated by varying the compositions of inventions *ad infinitum*. The idea is to determine the composition for which this minimum score turns out to be the highest (*Max)* and therefore the least damaging.

This cautious or "prudent" intervention (*MaxMin*) is the one with the highest possible GMiS level. It corresponds to the "value of play" in game theory.

The way in which the prudent intervention and the corresponding value of play are determined provides decision-makers with a sound basis for making decisions. Choosing the prudent composite intervention ensures that any possible future gives the company at least as much or more satisfaction with the GMiS. The strategic manager thus protects himself from the unexpected and is in a position to expect only good surprises in relation to the value of the game (GMiS).

Certainly, there are other interventions that, for certain scenarios, may be more satisfactory than the so-called "prudent" intervention. However, this advantage is sometimes illusory, because in certain contexts or particular scenarios, it risks being overemphasized in favor of the cautious intervention.

Above the GMiS, no firm, unconditional commitment to a level of satisfaction can reasonably be made without risk.

As an indication, the categories of results concerned by this cautious approach are:

- the execution of the intervention according to actual situations;
- surprises in relation to the expected situations;
- failures with regard to extreme situations (regrets, being overwhelmed).

6.1.4.2. *Other types of prudent strategies*

The *first type* of cautious or prudent strategy concerns strategies formulated on the basis of surprise. The example of stock market quotations reveals that behavior is more sensitive to the discrepancies between announced results and previous expectations than to the absolute value of the loss or profit made. In other words, the sanction is generally more severe when a profit is announced that is lower than expected, than when a substantial loss is anticipated.

The *second type* of prudent strategy deals with failures in extreme situations. These failures can be perceived in terms of:

– regrets related to the company's inability to exploit the best opportunities or ideal situations;

– overburdens due to their inability to avoid the worst threats or push-back situations.

Prudent strategies of the *third type* are based on the notion of regret. They represent the performance differential between, on the one hand, the ideal situation – that associated with the most favorable scenario – and, on the other hand, the actual situation. From this decision-making perspective, prudent intervention offers a "guaranteed minimum approximation" (GMA) of the best possible opportunities.

The *fourth type* concerns strategies based on measuring the overwhelming sense of the worst threats. Measuring the sense of something being overwhelming is the performance differential between the actual situation and the most execrable situation possible (push-back scenario).

In these four types of cases, characterized by an unweightable environment (that is, one where no weighting of the scenarios can be envisaged *a priori*), the possible future outcomes represent an infinity of plausible situations, associated with simple or remixed exploratory scenarios.

The so-called "prudent" composite intervention, as considered in game theory[3] (*Min/Max* overlaps), corresponds to the best "guaranteed minimum distance" (GMD) from push-back situations. The "performance differential" on which the GMD is based is defined as the mathematical expectation, that is, the average of the distances estimated within the framework of the different scenarios, with this average being

3 Game theory earned a Nobel Prize in Economics in 1994, won by three research authors: John F. Nash, John C. Harsanyi and Reinhard Selten. Nash's fundamental contribution was his concept of equilibrium, which we call "Nash equilibrium". This refers to "Any combination of strategies – one per player – such that no player regrets his or her choice after seeing the other players' choices" (Nash 1950b).

weighted by the subjective probabilities (of the variables and not of the data) that the "remixed" scenarios will occur.

Opting for cautious intervention ensures that any possible future will provide an average distance from the situation that is at least equal to, or even greater than, the GMD, so that the decision-maker can bear the burden. Certain interventions, within the framework of particular scenarios, may prove to be less overwhelming and more satisfactory overall than the so-called "prudent" intervention. However, these interventions are likely to prove – through a repulsive distancing from the GMD – more overwhelming in some future contexts. Thus, below the GMD, no firm (unconditional) commitment to a level of burden can reasonably be made without the risk of being misguided or even defeated.

6.1.4.3. *Justification and methods for applying the precautionary principle*

What can be said about the question of the "uses and abuses" of the precautionary principle? This section is an attempt to provide an operative answer to this question. Its purpose is to provide a formalized specification of the extent to which rational decisions concerning the application of the precautionary principle can prove to be erroneous, and lead – sometimes "in good faith" and in a consensual manner – to abuses of the principle.

The precautionary principle, succinctly presented in Box 6.2, is a rule for behavior and management in the face of an uncertain, "unweightable" environment.

> The precautionary principle has origins that link it to the concept of sustainable development. It has since been extended to many other fields: food, health, nuclear energy, climate, etc.
>
> This principle is frequently invoked, often without knowing what it involves; hence the need to treat it seriously in order to analyze its definition and clarify its content. The *precaution* must be based on the respect of rigorous procedures, which must be made explicit in a transparent manner in order to avoid the mistrust of citizens towards institutions that make them take ill-considered risks (of which they are not aware). Any action must, where appropriate, be accompanied by instructions for use, understood by all and of use to all actors.
>
> There are two approaches to risk: *prevention*, linked to the observation of a quantified reality of proven risks (based on objective probabilities of events), within the framework of a relatively curative approach; and *precaution*, relating to potential or hypothetical risks within the framework of an anticipatory and prospective approach. It is a perception measured in terms of the mathematical expectation of utility (based on subjective probabilities of possible situations or scenarios) from a long-term perspective and in a context of major uncertainties.

Insurance is concerned with the *prevention* of relatively externally generated risks (natural hazards, etc.), while *precaution* and foresight are concerned with internally generated risks, the damage of which can be attributed to possible perpetrators. Precaution consists in preventing the unpredictable, which it is difficult to say will not occur.

The precautionary principle must correspond to a new attitude towards risks, a form of "eco-citizenship" that refers to the notion of "precautionary ethics". The ethics of precaution is ambivalent. Indeed, it has a proactive, rather *visionary* aspect, which leads us to go beyond our limits so as to fight to prevent unacceptable damage from being done; a pre-active, more normative, or *conservative* aspect, through the reminder to take action, suggesting abstention. The concept of precaution obliges us not only to ask ourselves about risk (or its value), but also about the risk associated with this risk or "metamorphosis" (the value of its value). Should the lack of "proactive" precaution (morality) or the abuse of "pre-active" precaution (law) be sanctioned?

The exercise of precaution is considered a "noble" political act and relates to the methods of development and the conditions of acceptability of the residual risk, knowing that a case of zero risk does not exist. It is not a question of reducing the risk to the lowest possible level, whatever the cost.

Box 6.2. *Some thoughts on the precautionary principle (Ewald* et al. *2001; Kourilsky 2001)*

The precautionary principle is less of a principle of abstention, in the spirit of the saying "when in doubt, forbear", than a principle of action, whose slogan would be "when in doubt, do everything possible (for example, by detecting the right mechanisms) to act in the best possible way (by minimizing risk-taking)". This is why the approach that we propose to answer this delicate question revolves around an operational and original deepening of game theory, concerning prudent strategies for dealing with the worst threats.

This approach will attempt to clarify and explain the trade-offs most often made in application of the precautionary principle, in order to avoid falling into the trap of making emotional or political decisions in this area.

In light of the analysis elements from section 6.1.4.2 concerning the meaning of the conservative solution in terms of "guaranteed minimum distance" from push-back situations, we may then consider two cases.

Case 1: the tolerance threshold of what is overwhelming is lower than the GMD for push-back situations

In this hypothesis, the degree of fallibility of the intervention is acceptable and the application of the precautionary principle is not necessary. At this point, it is worth drawing attention to the fact that, if the prudent burden position and the corresponding level of GMD have not been determined, the political decision-maker may sometimes be led to abandon strategic interventions, making abusive use of the precautionary principle. This sometimes leads him or her to turn away, wrongly and irretrievably, from crucial opportunities for the future.

Case 2: the tolerance threshold for the overwhelming is higher than the GMD

Such a situation calls for the in-depth study of the following:

– the determination of the worst possible situation, corresponding to the GMD;

– the identification of other situations or scenarios corresponding to intolerable degrees of failure and strain.

An arbitration may then be rendered in favor of one of the following decision-making options:

– the disengagement or circumstantial abandonment of the intervention through the strict and legitimate application of the precautionary principle;

– a deliberate but reasoned intervention. In this hypothesis, it must be accompanied by an imperative of transparency on the part of the decision-makers concerning the real risks of default incurred;

– a (proactive) inflection of intervention conditions in order to prevent and eliminate the risks of fallibility.

6.1.5. *Changing the conditions to modify or transform the intervention*

Three mechanisms of change will be given here for purposes of demonstration. These are not exhaustive, but they can be combined to provide a range of proactive strategies.

6.1.5.1. *Reshaping or inflection of the maneuvering space*

The first mechanism change is to reshape or inflect the maneuvering space. The aim is to identify the constraints and requirements that hinder the pursuit of a given ambition and to measure the additional performance that could result from relaxing or bypassing them. Decision-makers then receive follow-up recommendations, in the form of rules of conduct, to overcome obstacles and "push the envelope of possibilities".

As an example, the remodeling or inflection of the maneuvering space may involve:

– the imposition of an untimely recommendation, not feasible a *priori*. In this case, the objective is not to adjust the intervention to the constraints, but rather to try to influence these constraints to ensure that its action is sufficiently viable;

– the resistance to prohibitive constraints so as not to be overwhelmed by a maneuvering space that could put the company out of business;

– the early removal of chronic or generic barriers that are persistent and pervasive, regardless of the options considered. The fact of pushing back such obstacles by anticipation may guarantee an improvement of the primary goal, even if it has not yet been chosen;

– the opening of loopholes towards ambitions of another order, which can be made concrete in terms of, for example, axes of breakthroughs, voluntary modification of the rules of the game, etc.

6.1.5.2. *Creation of the conditions to bring about a desirable future*

The second mechanism of proactive change is to invent a desirable future and to work to create the conditions to make it a reality. This ensures that what is desirable becomes possible and achievable. In keeping with Eleanor Roosevelt's quote, "the future belongs to those who believe in the beauty of their dreams", it is important to ensure that the dream of the company, proclaimed through its vision of the future, fertilizes the reality of its strategic context. To do so, the company must take the initiative to influence the course of events in such a way as to make the most desirable scenario happen, or at least to influence the likelihood of the environmental scenarios that work in its favor.

6.1.5.3. *Recasting of certain constraints and rules in force*

The third mechanism of change consists of an internal overhaul in the light of the constraints and rules in force. The organization will have to adopt measures intended to increase its productivity and competitiveness, in order to stimulate its vitality and increase its performance potential. It is a question of acquiring the capacity to "do more with just as much, or even with less".

6.2. Interpreting optimized solutions to support the decision

6.2.1. *Visualizing the results in a triptych of dashboards*

We can use an integrated triptych of interactive dashboards to represent the situation-problem and its solutions. The integrated nature of the triptych owes to the fact that a composite solution or intervention results in a morphological profile drawn simultaneously on all the different dashboards. The grids displaying the constraints and requirements of the situation form the background map of each of the three dashboards. These are presented in the form of "validity fringes" corresponding to the envelope of realistic solution profiles.

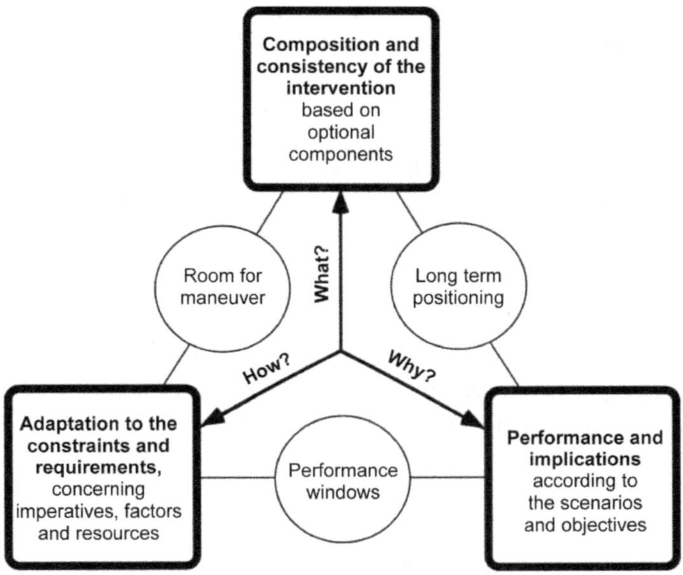

Figure 6.1. *A triptych of integrated dashboards duly completed for each decision recommendation*

The permanent refreshing of the results and the updating of these results allows for a monitoring-evaluation of the evolving situation and the resulting choices.

6.2.1.1. *Actual or potential performance*

The first dashboard involves the performance and implications of the solutions according to the goals or criteria considered. The background of this table shows, for each goal, the range of performance and the true potential it holds. This dashboard provides an answer to the question of the rationale for solutions: "why is it preferable?"

6.2.1.2. *Effective composition and consistency*

The second dashboard involves the composition and consistency of the composite solution. The background shows the true room for maneuver in the composition of the interventions. This table answers the question of the profile of the solution based on the components of the intervention: "what should be done?"

6.2.1.3. *Compliance with the system of constraints*

The third dashboard indicates the degree to which these solutions comply with constraints and requirements of all kinds (coupling of solutions, imposed thresholds, levels of available resources, etc.). The background map of this table indicates in particular the malleability of the constraints across the fringe of their boundaries. This last table sheds light on the question regarding how solutions are qualified with regard to constraints: "how is this possible and realistic?"

The interconnected nature of these dashboards is due to the fact that they offer the possibility of presenting and visualizing the morphological profile of any recommendation or solution according to their three optics, namely the composition of strategic tracks, the performances for the various objectives and the adaptation to the constraints.

These dashboards allow the "ultrasound" of the situation to be addressed, or the problem to be solved, and provide unprecedented keys for situational awareness and decision-making intelligence by understanding the geometric envelope of the solution profiles through the plurality of its facets.

More specifically, these dashboards help to reveal the true spaces of freedom and encourage the user to reshape them wisely:

– by delineating and tagging the windows of performance, capacities for action, and areas of possibility;

– by drawing up performance, activity, and resources profiles;

– by indicating the procedure to be followed to favorably modify the contours;

– by showing, in real time, the modification of intervention profiles or the remodeling of spaces of freedom.

6.2.2. *Revealing performance levels for the objectives-criteria*

The first dashboard allows us to view the envelope of possible consequences to be shown according to the different scenarios, that is, the impacts on each objective of any composition of strategic initiatives that respects the constraints. The knowledge and consideration of the constraints are translated into successive ranges of scores for each of the objective scenarios.

These ranges represent the windows of satisfaction that the company could achieve in each scenario, regardless of the composite course of action (or the mix of initiatives) it is considering.

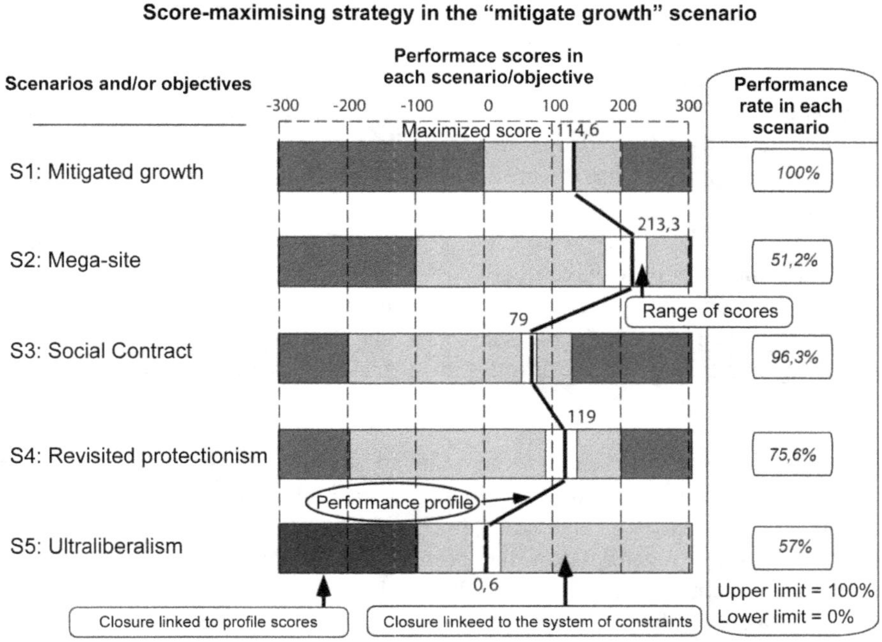

Figure 6.2. *Performance dashboard and implications for scenarios and/or objectives*

This gives the management team the opportunity to visualize and appreciate, through the lens of its concerns and requirements, the interest that can be taken in the pursuit of certain scenarios through the establishment of a "hierarchy of intervals", according to which these scenarios will potentially prove to be more favorable or more discriminating. In particular, this hierarchy makes it possible to internalize the constraints and to clarify the role played by them, by highlighting the performance that the company could achieve in the context of the various scenarios (or with regard to the objectives pursued). This underlying method, which consists of minimizing and then maximizing the scores of the composite action tracks for each scenario-objective, has proven to be more efficient than a classical exploration procedure based on a large number of simulations.

This additional feature of optimization, allowing for considerable savings in the problem-solving process, regardless of the recommendation of particular solutions, would seem to be decisive.

Thus, all strategic paths can be represented schematically in the form of a specific score profile. This profile, provided that the strategic path in question respects the system of constraints, is located within a snake-shaped zone covering these ranges. However, it is not sufficient to sketch a profile within the snake outline to determine that it corresponds to a composite strategy admissible with respect to the constraints. In particular, the profile of ideal performance through the ceilings of all scenarios or targets is generally not realistic. Thus, the effectiveness of the system of constraints is expressed, beyond the envelope of possible performance profiles (in the form of a snake in the dashboard), by presenting these profiles in a zigzag form, adhering to a certain logic. The optimization according to a given scenario or objective – which consists in imposing the passage of the profile through the ceiling point of the range in question – makes it possible to vary the entire profile in terms of performance.

Such a schematic presentation amounts to showing ranges of possible score variations and makes it possible to calibrate, for each of the scenarios/objectives, the assessment of its results, expressed in the form of an effective rate of performance. These rates vary, regardless of the scenarios/objectives, from 0% (where the performance is equal to the floor) to 100% (where the performance is equal to the ceiling). Evaluating the degrees of performance of any strategic composition in the form of percentages also offers the possibility of carrying out multiscenario or multiobjective optimizations. This feature is of particular interest when the performance scores according to scenarios or objectives are expressed in different units.

6.2.3. *Describing the composition of the options and exploiting the room for maneuver*

The second dashboard displays the composite intervention (recommended mix of actions) via the profile of its relative weights or levels of use of its various components.

Regarding the components that should be excluded from the optimal composition, the dashboard presented in Figure 6.3 provides a comparative indication of the handicaps assigned to each of the components considered as potential candidates. This acts as a detriment – expressed in terms of the deterioration of the score obtained for the optimized objective or scenario – that could result from their forced inclusion in the composition. Thus, the knowledge of these *combined handicaps* to selection allows the company to have a hierarchy between the excluded components. This can be particularly useful in the case of negotiation or compromise research.

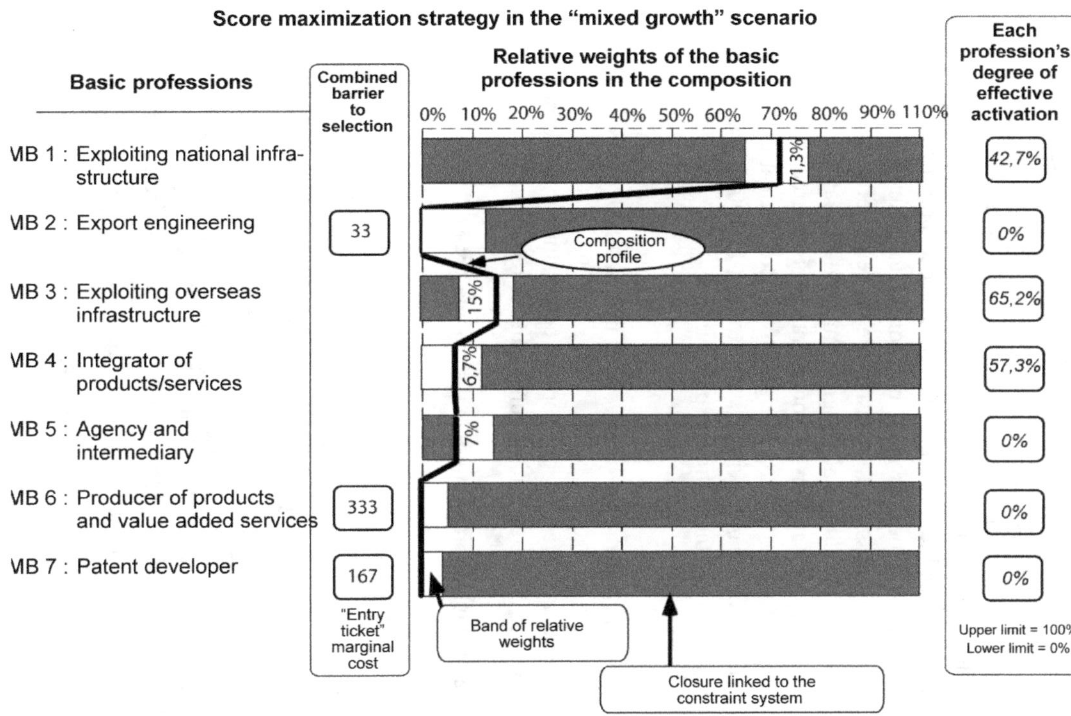

Figure 6.3. *Composition of the company's business based on the lines of its core business*

This dashboard dedicated to the composition of solutions includes the ranges of the relative weights of the possible elements. Thus, although these weights are supposed to be between 0% (excluded from the composition) and 100% (a single component), this *theoretical* window can be narrowed down, resulting in an *effective* window. Such a window would in turn be capable of being scanned by the relative weights of the component concerned, in the context of cohesive compositions, respecting the system of constraints. Thus, revealing the floor and ceiling weightings associated with each component of the strategy (in other words, the true room for maneuver) constitutes valuable information for the decision-maker. It offers a panorama of the real possibilities for each component to contribute to a coherent strategic profile. It makes it possible to measure the degree of effective activation of this component, in terms of additional weight compared to that which would – in any case – be imposed by the system of constraints. This degree is measured as a percentage, from 0%, when the relative weight of the component does not exceed the floor attributable to compliance with the constraints, to 100%, when it reaches, within a realistic composition, the ceiling of possibilities.

6.2.4. *Identifying saturation and estimating the pressure of constraints*

The third dashboard involves the pressure of constraints or more precisely the adaptation of the solution to the system of constraints or requirements.

Score maximization strategy in the "mixed growth" scenario

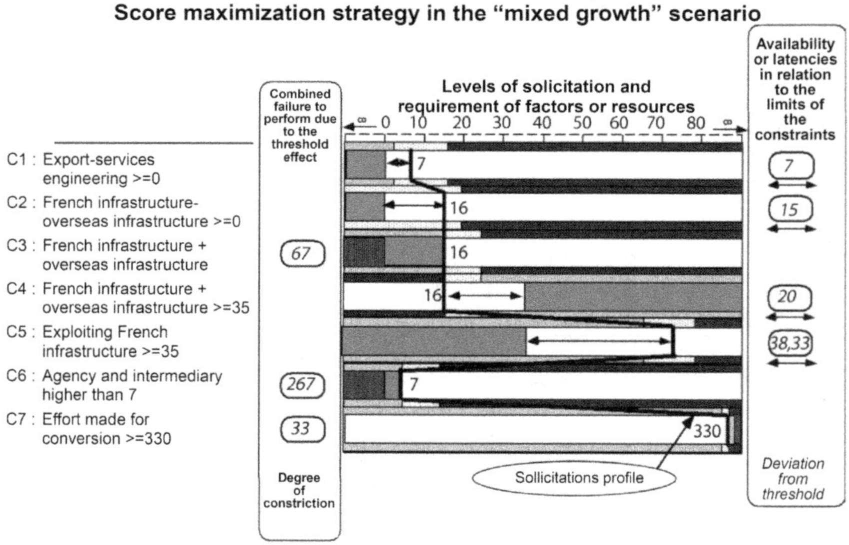

Figure 6.4. *Adapting to constraints and requirements: strategy score maximization in the "mixed growth" scenario*

In this dashboard, each of the constraints is expressed by the validity domain of the values of its first member (mathematical expression) in relation to its boundary (second member). Thus, any composition of strategic tracks is translated by the drawing of a specific profile, crossing the validity domains of each of the constraints. In addition, the calculated values of the first members of the constraints are presented, as well as their deviations from the boundaries.

These deviations express the potentialities or latency levels offered by the constraints, respectively. Each latency denotes, in fact, a certain potential flexibility capacity of the firm in relation to the constraint concerned. Moreover, since these are saturated constraints (for which the limits are effectively reached), the dashboard provides additional information in terms of the degree of effective pressure of the constraints.

These degrees allow for the prioritizing of the level of "effectiveness" of constraints as bottlenecks, with their value being zero for unsaturated or loose constraints. These degrees result from a judicious appreciation (the calculation of dual variables) of the way in which the saturation effect of the constraint affects the maximization of performance for a given scenario or objective. In other words, it expresses the price of scarcity or the opportunity cost of the factor concerned by the constraint. This allows us to evaluate, almost automatically, the opportunity to incorporate any new strategic approach to any optimized composition. This is to say, this dual variable provides the extent to which the maximum level of performance could – for the scenario considered or for the objective pursued – prove even better, possibly by relaxing the constraint via a unit-level shift of its boundary. This gives rise to the term "unit shortfall in performance" on the scenario/objective, due to the saturation of the constraint.

This valuable information helps to identify the constraints or requirements that "hamper" a given objective and to indicate the extra performance that could result from relaxing them. It is important, in this case, to no longer persist in making the most of the straitjacket of current (responsiveness) or anticipated (pre-activity) constraints and requirements by optimizing its performance. The idea is then to adopt a rule of conduct or imposing detailed recommendations in an attempt to reshape the field of possibilities (proactivity), with the goal of surpassing this record performance. The company will then have to guard against optical illusions by detecting the real obstacles, as well as the "relevant breakthrough axes" to overcome them.

This gradation in terms of the strategic issues consists of going beyond a reactive or pre-active attitude in order to reach an eminently proactive posture. To be more complete and fully operational, this rule of conduct must also indicate the limit to be respected in order to loosen each constraint by moving its milestones.

The threshold in the direction of tightening is called the "incompatibility threshold" of the constraint system. It marks the limit of movement of the constraint's stop-block, beyond which the problem does not allow for a solution (with all constraints otherwise unchanged).

The other threshold, known as the "redundancy threshold", expresses the limit to which the constraint's upper bound can be loosened, beyond which the constraint is satisfied regardless of the scenario in question, the optimized objectives, or the recommended solutions. This constraint is therefore redundant with respect to all the others.

Finally, for any optimized objective, the sensitivity analysis of the constraints provides a sub-range around the milestone, within which any movement of the stop does not lead to a change in the choice of the strategic tracks of the optimal composition. The relative weights are then modified. This band constitutes a zone of relative stability of the solution, which the company must know how to take into account, avoiding discontinuities at the borders, in order to hope to control the change, especially when it proves to be voluntary. Thus, this dashboard offers the decision-maker an exceptional opportunity to visualize (using the form of Russian dolls) a series of maneuvering areas constantly increasing in scope, depending on the degree of proactivity they require from the company. During our presentation of the successive uses of the results, it was noted that the role of the optimization function progressively disappeared, replaced by a better legibility of the constraints.

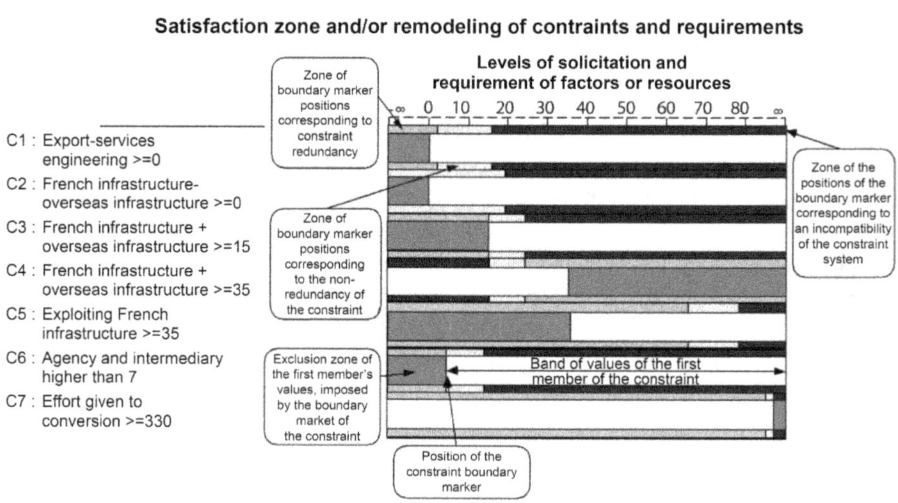

Figure 6.5. *Adaptation to the constraints and requirements regarding core businesses, factors, and resources*

First, the multicriteria optimization function was limited to implicitly integrating the constraints. Then, the circumstantial fitting of the constraints to functionality was accompanied by highlighting the only constraints that were not respected. Finally, the legibility of the constraints becomes overwhelmingly clear, thanks to their visualization and their internalization by decision-makers on different registers (effective performance ranges, real room for maneuver, prediction of constraints, etc.). These registers prove to be closer to the real concerns of the company.

6.2.5. *Detecting and estimating the value of proactive deposits*

This last use of the tool can also induce proactive actions, allowing the "envelope" of obstacles and constraints to be "pushed" to access a previously unthinkable "breakthrough area".

Thus, being aware of constraints and their true role transcends the search for solutions of a pre-active nature, by inscribing the constraints within more meaningful spaces. New, more proactive, and therefore eminently strategic challenges can therefore be envisaged. This functionality, which opens access to a higher order of complexity, is more intelligible for most decision-makers.

The systematization of this *hyperfunctionality* of the tool opens a potentially beneficial path toward deepening our research in the praxeological field. In particular, it could enrich the approaches to understanding and modeling the situation problems (or *problematization*) with which companies and organizations are faced. This enrichment, heuristically speaking, would consist of shaping the representation of the problem through a progressive back-and-forth between the exploration of solutions and the morphology of the situation.

PART 3

Acting and Intervening: Judiciously Guiding Action to Generate a Virtuous Dynamic

7

Identifying Predispositions to Act and Generating Compromises

As a result of our attempts to formalize concrete systems of relationships, both intersubjective (peer-to-peer) and interorganizational, in order to better interpret them, we have developed a pragmatic concept intended for the analysis of tables of interactions between the most important players, for praxeological and strategic purposes. This concept, inspired by the notion of the "composite exchange" put forward by François Perroux (1981), initially led to the publication of an article (Arcade 1990). Following this, it gave rise to the development of a software tool, Astris (Arcade 1990) (see Box 7.1).

Astris is an instrument for performing systemic analyses of tangled sociopolitical situations. It allows for actors' games to be deciphered, in order to manage dynamics of compromise and thus take full advantage of them.

More specifically, this software works by mapping the interactions of the actors in order to analyze their power relationships and their balance of power, decrypt their behaviors, and diagnose the points of convergence or "apples of discord". In addition, by offering a "panoptic" view of this *game* and the analysis of its distortions, it makes it possible to provide early warnings and to regulate their change over time in a beneficial way.

The use of Astris is particularly productive in areas such as sociopolitical conflicts, geopolitical balances, exchanges of goods and services, tangible and intangible communication networks, cross-participation and capital transfers, socio-organizational cohesion and convergence, and the evolution of electoral landscapes.

Box 7.1. *Astris: strategic analysis of systemic interactions*

This innovative instrument for analyzing networks of exchange or interdependencies combines theoretical and applied propositions, including game theory, multicriteria analysis, index theory, multivariate utility functions, and the notion of structural drifts. In addition, the instrument-based approach discussed in this chapter has given rise to various applications or experiments. It is an original way to deepen, broaden, and even go beyond the methods of deciphering transactions between actors and discerning their games and stakes, allowing for the possibility to of putting in place a virtuous dynamic of "cooperative" transactions. It also offers a series of key concepts for the judicious interpretation of the intertwining of periodic interactions, using a palette of heuristic maps.

This chapter includes two types of dashboards, on the one hand, conceptual or methodological diagrams, and on the other hand, heuristic maps or grids, which can be interpreted relatively intuitively, as a result of the projection of variable points in appropriate planes. This second type of graph is to be preferred for an immediate and practical application of the method.

7.1. Deciphering the power relationships and power factors of the actors

The base of support on which the "composite exchange" rests can be found in the communication and influence graphs and constitutes the substrate of the system of relations. Beyond its technical aspects, the conceptual dimension becomes fundamental in detecting and apprehending certain complex and intertwined phenomena involving forces, dominances, power, and the convergence of behaviors. The Astris tool is therefore based on the identification of relational or organizational networks. It allows for an analysis of the interactions within nested systems based on the modeling of the strategic landscape at different scales, as revealed from our engineering approach. The underlying principle is that of remaining unsatisfied with superficial observations of the tangible flows of exchanges through identified networks of relations, instead focusing on their symbolic expression. This essentially praxeological expression is in fact based on the notions of power and power relations that emerge from "cross-utility functions".

We make use of a *matrix of ambivalent synergistic interactions or antagonisms between actors*. All such interactions are qualified and assessed within the systemic modeling of the strategic landscape, according to an appropriate scale of attributes of the following type, in which certain scales of synergy or antagonism can be suppressed:

– synergy:

- strong,

- somewhat strong,
- somewhat weak,
- weak,

– neutral (neither synergy nor antagonism);

– antagonism:

- weak,
- somewhat weak,
- somewhat strong,
- strong.

This matrix can be transformed into a *matrix of net synergies* (synergistic interactions) based on the difference (in absolute values) between synergies (positive) and antagonisms (negative). These net synergies can be translated so that they are all positive. In this case, which is the default rating method, the qualitative scale of attributes presented above, as an example, would correspond to a four-point scale of ratings ranging from 1 (strong antagonism) to 9 (strong synergy), with a mid-point of 5 (neutrality).

The initial matrix could be transformed into a *matrix of net antagonisms,* which in this case would result in an inversion and translation of the attribute scale presented above. It is worth mentioning that the exclusive use of one of the net synergy or antagonism matrices would imply giving precedent to a particular focus on one of these two monovalent aspects. By mitigating or even "erasing" the disruptive power associated with antagonizing detractors, this approach constitutes a *de facto* "hemiplegic" analysis of the synergies and/or antagonisms between actors. It therefore requires that particular attention is paid to the resulting interpretations of actors in the opposite hemisphere.

Box 7.2. *Some observations on the assessment of interaction scores*

7.1.1. *Schematizing the conceptual approach and introducing key concepts*

7.1.1.1. *A triptych of analyses*

The approach used in Astris is based on a triptych of analyses – partially static globally static, and dynamic analysis – which will be examined below. Each concept or notion allows for enlightening results and measurable indicators to be obtained.

These analyses concern the processing of temporal series of cross-tabulations (or double entry matrices) of actors and objectives. Figure 7.1 illustrates each type of analysis, shown within ellipses. The subsets of tables submitted for processing are specified by hatching, and the methodological bridges between these types of analysis are shown.

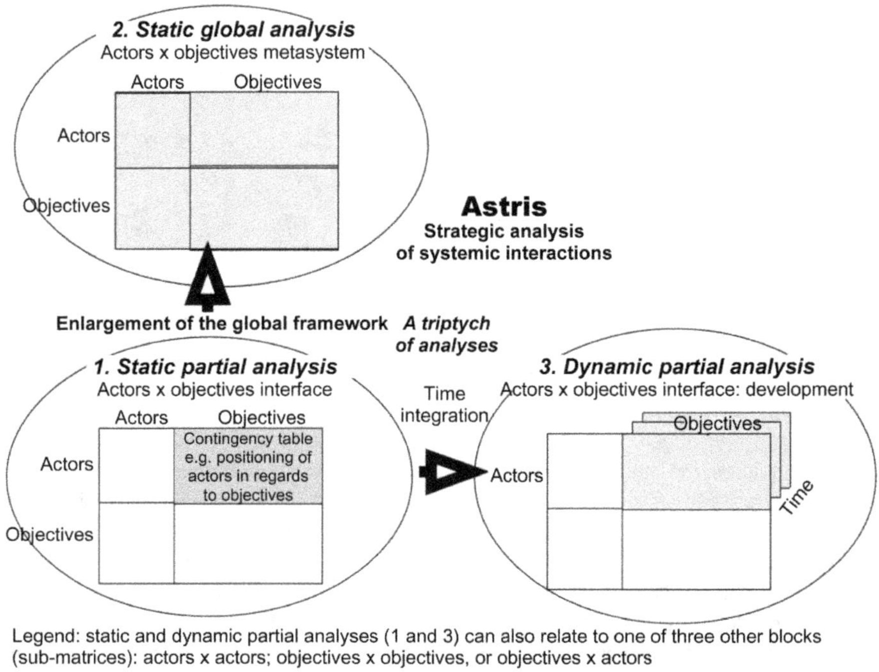

Figure 7.1. *Strategic analysis of systemic interactions: connecting the approach*

The three mechanisms of analysis (partial, static and dynamic) shown in Figure 7.1 are described briefly as follows:

1) *Partial static analysis* is limited to the observing of particular subsystems (i.e. actors or objectives) or their interfaces. Based on the notions of power and power relationships, it provides, a key for visualizing and deciphering the situation as it is observed. This key highlights the articulation between the posture and stature of the actors and the identification of dissonances or cognitive consonances between them.

2) *Global static analysis* concerns the (meta)system that integrates several subsystems and includes their interactions. It allows for the evaluation of the existence, or internal power, of an actor as being constituted by an accumulation of strata, from its intrinsic power to the effects of vacuity induced by its relations to the system (material exchanges or sharing of affinities).

3) *Dynamic analysis* concerns the deformation of the subsystems or their interface between them (partial analysis). It allows us to define and measure dynamic excessive power, which is to say, the overvaluation of the power of a player, or its "access to power", as resulting from the dialectical overcoming of aggressive initiatives and the system's carelessness.

For pedagogical reasons, our successive presentation of the functionalities of Astris for the different areas of the analysis triptych is limited to applications concerning the positions of actors in relation to objectives. However, the generic character and the structuring mode of the Astris concept offers the possibility of switching from the *primal* approach (e.g. in terms of actors) to the *dual* approach (e.g. in terms of objectives) and vice versa.

7.1.1.2. *Direct or induced modes of obtaining results*

All results concerning the actors can be obtained:

– directly from the matrix (diagonal block of the strategic landscape) of interactions between actors;

– indirectly via the matrix of the *actors' influences* or *hold over the objectives*.

This last calculation mode is the basis for our presentation. In the same way, all of the results concerning the objectives can be calculated:

– directly from the (diagonal) matrix of interdependencies between objectives;

– indirectly through the (non-diagonal) matrix of influences of the objectives on the actors.

7.1.1.3. *The push–pull optics of analysis*

Each of the types of analysis performed by Astris is made using two contrasting and inseparable *push* and *pull* optics. In the push optic, the actors are considered as transmitters of influence. Conversely, their role as receivers of influence is prevalent in the pull optics.

The comparison of the different results and indicators arising from the push or pull optics makes it possible to interpret and visualize the ambivalence of the role of each actor in the system of interactions in which it participates.

These two perspectives are taken into account in all calculations and in most of the visual presentation of the results generated by Astris. Thus, in addition to the edition of graphs tracking each of the two optics, the software tool offers the dual possibility of superimposing the resulting graphs. This tool also enables the creation of an additional graph, in which the coordinates of the points from these two optics are replaced by their push/pull ratios.

In the particular case where the matrix concerns the relations of exchange or influence between actors or within a set of entities of the same nature, the singular and dual approaches take on a particular meaning. They both concern the same set, but differ in that the singular approach concerns the pushing aspect of the actors, considered as emitters of influence or as origins of flows. The transposed approach, which is symmetrical, refers to the pull aspect, in which the quality of influence receivers or recipients of flows is emphasized.

Although our purpose is limited to essentially pedagogical purposes, thus to the interactions of actors, or the objectives of the strategic options under consideration, it can also be applied to other categories of elements in systemic interaction, such as the components of the strategic landscape, like the hypotheses of the scenarios, the interinfluences between actors and scenarios, the interdependencies between the scenarios, the orientations of the strategic frame of reference, etc.

7.1.1.4. *The notions of cross-utility, self-satisfaction, and ascendancy*

This initial analysis allows us to validate the articulation between the observation of exchange flows (operating level) and the higher order symbolic expression (meta level) of power relations[1] and power relationships. This type of connection is based in particular on the demonstration of an equivalence between the weight of the competitors in the exchange and the equilibrium of a strategic duel (in the sense of game theory) based on a matrix of “crossed utilities”. The methodological pathway to support this fundamental equivalence is shown in Figure 7.2.

1 According to François Perroux (1973), “The exchange is first and foremost inseparable from the concept of power, which would appear – beyond the simple accounting work done on prices and quantities – in the form of transfers of influence between suppliers, considered as actors in their own right. This transfer takes place through asymmetrical interrelationships, the intensity of which stems from judiciously defined cross-utility functions”.

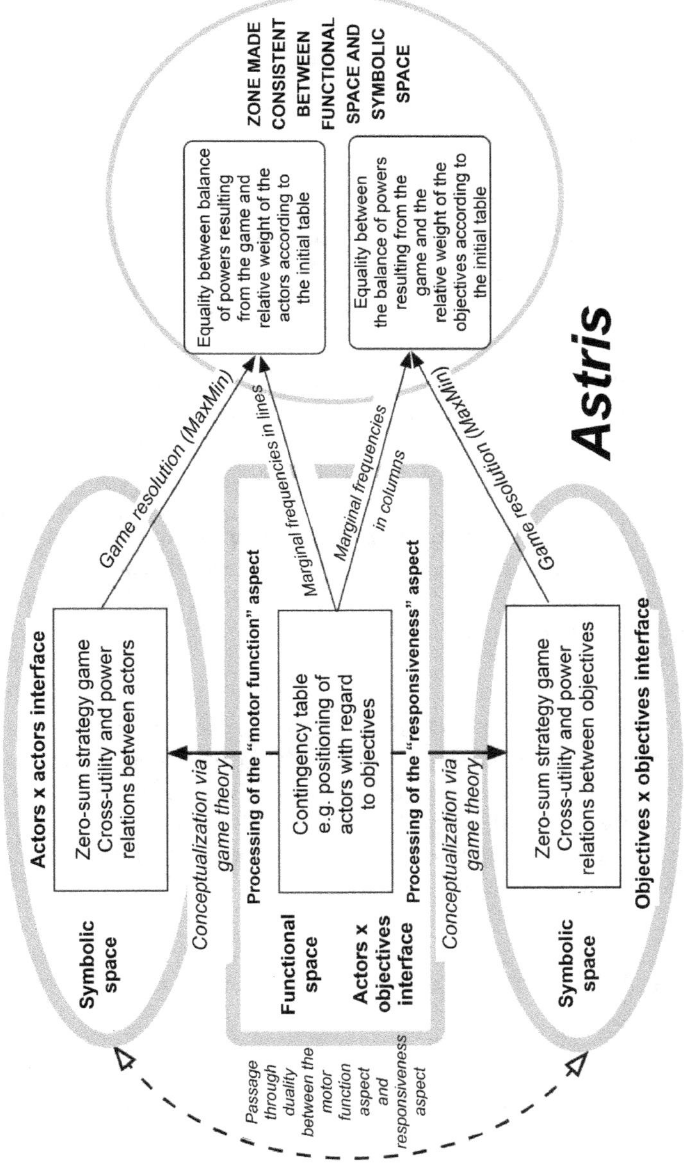

Figure 7.2. *Static partial analysis: conceptual map*

7.1.2. *Determining the power effect (or leverage)*

The cross-utilities, as evaluated by Astris, result in two sets of indicators reflecting:

– the satisfaction of the competitors concerning their own situation, or own utilities, corresponding to the level of "sufficiency" or strength;

– the affect of the actors, that is, their degree of attraction to others, through indices of each actor's preference for the situation of the various players.

7.1.2.1. *Characterization of relational strength, power, and power effect*

The relational strength, or the stature of a given actor, is equal to its own utility or its "intrinsic utility". This represents the diagonal element relating to this actor in the cross-utility matrix. The interrelational strength reflects the actor's stature. In other words, it reflects the actor's capacity to balance the efforts it makes on the various actors according to the importance they have in its eyes. It reflects the degree of existence or extrinsic presence of this actor through the accuracy of the "dosage" that it manages to achieve in its relationships with others.

The power of an actor or the effective weight it carries in the game corresponds to its relative weight in the exchange. The equivalence between the power and relative weight of the players refers, more precisely, to the equilibrium of a game or strategic "duel", the table of gains of which would be provided by the cross-utility matrix.

The multiplying coefficient, which makes it possible to go from the level of force of a given actor to the measure of its effective power, represents what is called the "power lever" or "power effect" of this actor. This notion reflects the actor's capacity to effectively mobilize its force or, in other words, its ability to "take action"; for it would be pointless for this actor to multiply the effects of an announcement concerning their presumed force if it is not able to put it to the test at the appropriate time.

7.1.2.2. *Visualization of the forces and powers of the actors and their ability to "act"*

The graph presented in Figure 7.3 allows us to analyze the power levers available to the different actors in the interaction. It is presented in the form of a plan named "power levers or power effects" with the relational strength on the x-axis and the effective power on the y-axis. A projection of the different points representing the actors is made on this plane. Thus, the multiplier coefficient translating the power lever of a given actor point corresponds to the "slope" of the oblique segment joining this point to the origin of the plane, so that the first bisector (diagonal southwest/

north-east) is the "geometric locus" of the actors, which are characterized by the unitary level of their power lever. On this basis, the actor points below this bisector have a power lever below unity. These actors can therefore only transform their relational strength to a limited extent, through the manifestation of an effective power of a lesser magnitude; this would make it difficult for these actors to take action. However, they would be inclined to compensate for this relative powerlessness by a certain penchant for "gesticulation". On the other hand, actors whose representative points lie above the diagonal benefit from a "power effect" (power lever) that is all the more favorable as they move away from the bisector.

The application of this analysis key to the whole plan allows us to derive an indicative typology of its different zones. The specific meanings of these zones, which are intuitive, were the subject of the unprecedented interpretation grid for the positioning of the actors presented in Figure 7.3. The distribution of these zones must be considered in a relative manner; this is in fact sub-determined by the general appearance and visual configuration of the actor point cloud projected in the plan.

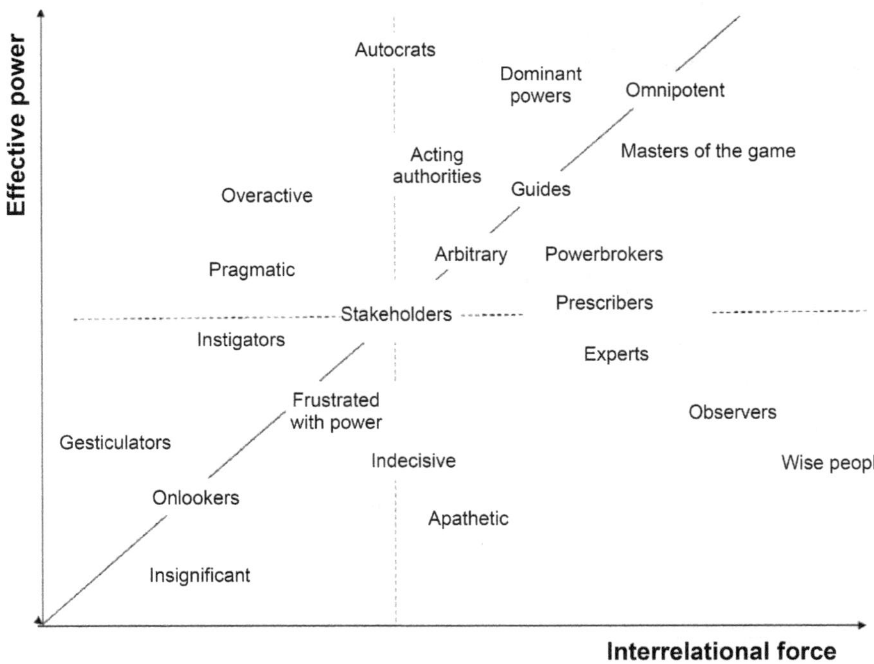

Figure 7.3. *Map of power levers or power effects: indicative grid for interpreting the position of the actors*

Figure 7.3 shows, among other things, that the "overactive" elements, though limited in strength, tend to overcompensate for this with a relative excess of actions, the effects of which (in terms of effective power) are apparently not enough to allow them to take a dominant position.

7.1.3. *Conducting a structural analysis of the actors in terms of their strengths or powers*

It is possible to obtain a graph that displays either forces or powers, allowing us to analyze the way in which the push and pull values of these factors are combined for the different actors of the interaction.

7.1.3.1 *Push/pull forces or power switchover graphs*

This so-called "force or power switchover" plan also represents a structural analysis graph of the tangled relations between players, in which the pushes (y-axis) refer to the notion of influence, and the pulls (x-axis) to dependence. The interpretation of the four main classes of variables resulting from a classical structural analysis is presented in Chapter 1 (section 1.2.2).

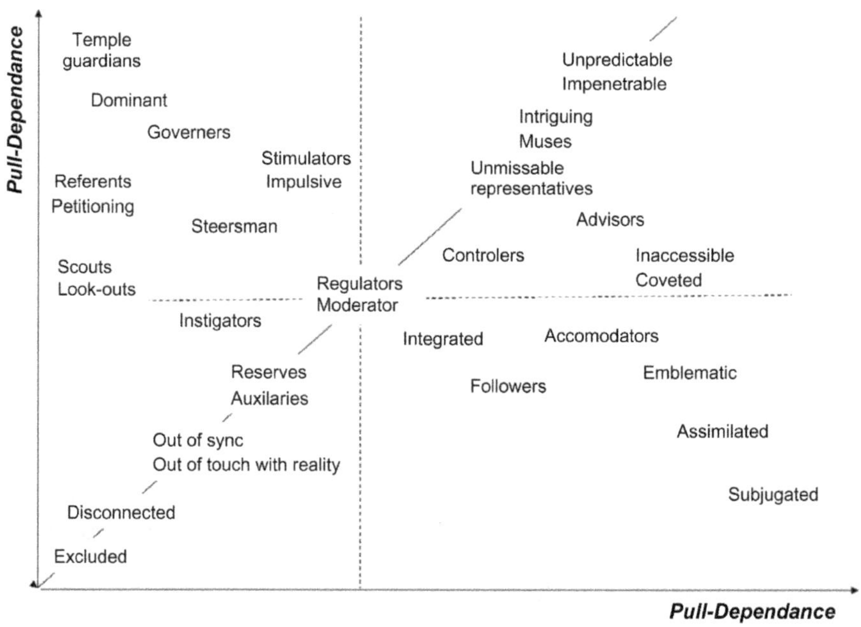

Figure 7.4. *Push/pull graph showing the switchover of forces or powers: indicative grid for interpreting the role of the actors*

For example, the graph of forces is presented in the form of a plane, in which the different points representing the actors are projected, with the push force on the y-axis and the pull force on the x-axis. In an analogy with the power levers graph, the key to interpreting this type of graph is very simple. The actors all have forces – both push and pull – that are higher the closer they are to the upper right corner of the plane, and vice versa.

In addition, the actors above the first bisecting line are able to amplify their force when switching from push to pull optics, especially as the line connecting them to the origin of the plane slopes upwards. Of course, the reverse of this interpretation applies in the case of the actors located below the bisector.

7.1.3.2. *Additional details concerning the two optics*

Based on the positioning of the points on the force or power switchover plane, it is important to refine both the distinctions and groupings of the actuator points. The interpretation grid (Figure 7.5, section 7.1.4.2) of the structural role of actors in the network in which they participate is the result of experience in effectively creating and interpreting this grid. In particular it is based on the fact that the distinction between push and pull optics is much more nuanced and subtle, as can be seen in Table 7.1.

Push optics	**Pull optics**
Influence *transmitter* *Impulse* role Propensity to *push* the game *Weighs on* the system	Influence *receiver* Role of *attraction* Propensity to *pull* the game Reflects the system or *takes advantage of it*
Influence: having a *hold on* the system Early/cause Given/condition/assumed/expected Prerequisites/requirements Lever/spring/stimulation Induction/propagation/diffusion Inertia/gravity	*Dependency*: impacted or conditioned by the system Involvement/events Symptom/effect/impact Incidence/outcome/ result Deduction/convergence/concentration
Influence factor/control *variable* Departure/origin/entry Origin/contribution/initiatives Source/home	*Status* indicator/effect *indicator* Propensity/trend/perspectives Arrival/end/exit Destination/assignment Target/focus

Table 7.1. *Semantic distinction between push and pull optics*

The richness of the meanings contained in these two notions (push/pull) requires special attention so as to encourage the gradual development of a valuable skill in strategic intelligence in delicate situations. For this reason, this interpretation grid is indicative, perfectible and pedagogical in nature (Figure 7.5, section 7.1.4.2).

7.1.4. *Interpreting the power relationships according to the stature and posture of the actors*

The balance of power between two actors is the quotient of their "intrinsic utilities". The actor of reference is the one whose strength acts as the divisor of this ratio, the other actor (as the dividend) is called the protagonist in this bilateral exchange.

7.1.4.1. *Breakdown of power relations into power factors*

This concept makes it possible to break down the balance of power into a product of two power factor, reflecting the dominances of power in terms of "stature" and "posture".

Stature represents the tangible power of the protagonist actor over the reference actor. It constitutes an ascendant attributable to the relative stature of the protagonist and generates a "Goliath" type effect.

On the other hand, posture is more a matter of subtle power that the protagonist exercises through its ability to take advantage of the situation and its positioning. This, by contrast, can be considered as a "David" type effect.

The Astris tool calculates bilateral ascending factors of posture and stature between actors based on "index theory"[2]. These ascendants are composite or synthetic indices, of the "Laspeyre" type in the case of stature, and of the "Paasche" type for posture. From a conceptual point of view, these two types of indices refer to the notions of quantity and price. On the other hand, power relations are analogous to elementary indices of value.

---

2 Index theory is built on the utility function of the consumer and on the hypothesis of the rationality of economic agents. It is the origin for evaluating GDP in volume terms, international comparisons of productivity, etc. Index theory is based on the utility function of the consumer and on the hypothesis that economic agents will act rationally. In the framework of Astris, we use this theory from a mathematical and computational point of view. In particular, we adhere to the criticisms formulated by Herbert Simon concerning the circumstances that can increase the risk of a lapse in rationality.

For a given reference actor, the ascending index is higher (or lower) than unity depending on whether it considers its own situation to be better (or worse) than the situation of the other actor. This index is therefore equal to the unit (base 1) in terms of its assessment of its own situation.

7.1.4.2. *Interpretation of power relationships in the posture/stature plane*

We may create a visual representation of the situation of the different protagonists, as perceived by the reference actor, by making a graphic representation through a posture/stature plan. The points representing the different protagonists have the values of the synthetic stature indices as the y-axis and those of the posture indices as the x-axis.

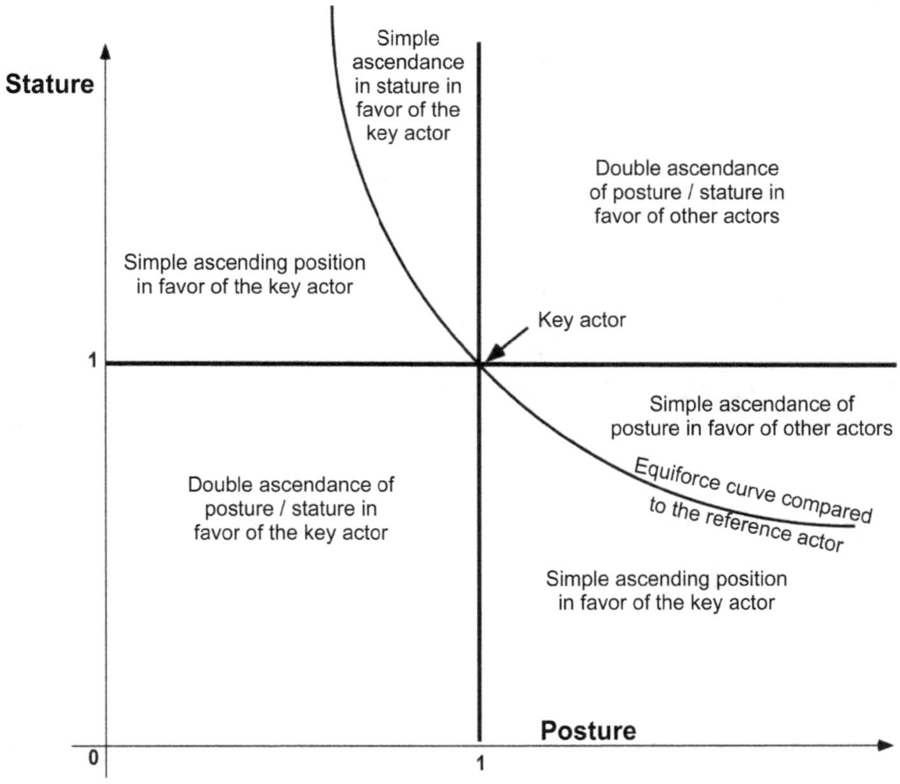

Figure 7.5. *Graph for interpreting the situation posture/stature of actors with reference to a particular actor*

The representative point of the reference actor is located at the intersection of the horizontal and vertical lines corresponding to unitary indices of stature and posture.

These lines allow the partition of the posture/stature plane into four quadrants, offering an interpretation grid of the actor-point cloud. This is based on the concordance and/or discordance of the posture/stature pair of the different actors in relation to the reference actor, depending on the reference actor's point of view.

The lower left and upper right quadrants represent the locations of the actor points for which, according to the reference actor, there is an unequivocal appreciation of the postures and statures of these other actors, compared to its own. The reference actor believes that it is in a strong dominance relationship, one similar to master and slave, due to its double advantage in terms of posture and stature, with respect to the actors-points in the lower left quadrant. On the other hand, this relation of domination is reversed (unfavorable) vis-à-vis the protagonists in the upper right quadrant.

The lower right and upper left quadrants refer to the protagonists for whom we note, with respect to the reference actor, a certain vagueness in the interpretation of posture and stature. This actor is in a mixed situation, characterized by both unfavorable stature and favorable posture in relation to the protagonists in the northwest quadrant. This relationship of ascendancy, of the chief collaborator type, is therefore rather unfavorable to them. However, this positioning offers the reference player an opportunity to play a role of "grey eminence", due to its better fit with the structure of the objectives. Symmetrically, the reference actor is, thanks to their better stature, in a position of leader vis-à-vis the protagonists represented in the southeast quadrant. They will therefore have to ensure that they are in a position of compose with regard to these protagonists, because of their advantageous posture. The Astris concept removes the ambiguity attached to the interpretation of the points of the southeast and northeast quadrants.

By construction and by definition, the balance of power is decomposed into the product of the two factors, respectively posture and stature. The set of protagonist points whose force ratio is balanced with respect to the reference actor is thus found on the curve in the form of an "equilateral hyperbola branch" drawn in Figure 7.5. This curve is of equal force in relation to the reference actor, dividing the posture/stature plane into two areas whose significance is determined through an initial analysis of the scatter plot. Indeed, all the protagonist's points found in the area located to the upper right of this curve benefit from a favorable power ratio with respect to the reference actor. On the other hand, this ratio is unfavorable for the protagonists located to the lower left of the hyperbola.

Thanks to this new partitioning of the posture/stature plane, it is possible to refine the analysis of the protagonists whose positioning has proven to be mixed. The protagonists belonging to the northwest quadrant observe a "simple ascending stature" relationship with the reference actor. This relationship corresponds to a ratio of strength that is favorable or unfavorable to them, depending on their position in relation to the curve including the reference actor. In the same way, the nature of the relations that the protagonists in the lower right quadrant maintain with the reference actor is of the "simple ascending posture" type, either in their favor or against them, depending on the part of the plan they occupy.

The subtle application of this key for analyzing the positions of the points of actors on the overall posture/stature plan has made it possible to adopt an indicative graph for interpreting the power relationships and power factors of the actors. As will be seen, this grid (Figure 7.6) is intuitive. It can be improved, and will have to be adjusted on a case-by-case basis according to the general appearance and configuration of the scatter plot of actors projected in this plan.

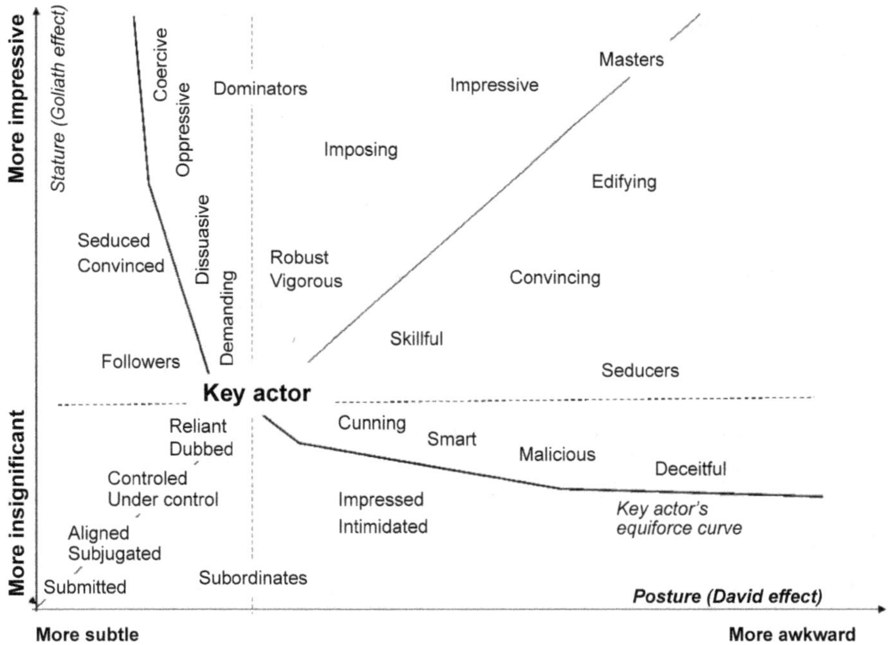

Figure 7.6. *Posture/stature graph for analyzing power relationships between actors: indicative grid for the interpretation of the power relationship and power factors of the protagonists with respect to the reference actor*

7.1.4.3. *Identification of cognitive consonance and dissonance*

The same analysis can be carried out from the point of view not of the reference actor, but of the other actors. In this case, the preference indices of the reference actor for the situation of the others will be based on the perceptions of the latter. We can then proceed to the projection of the actor points resulting from this new analysis on the same posture/stature plane. Astris then draws arrows connecting the points representing the posture/stature of the same actor, according to the perception of the reference actor (points from the first analysis) and that of the protagonist in question (points from the second analysis).

The interpretation of the double scatter plot of actors, represented by a field of arrows, makes it possible to highlight the convergence or divergence of appreciation or, in other words, the "cognitive consonance or dissonance" of the reference actor with any other protagonist regarding the situation (posture/stature). First of all, it should be observed that the points representative of the same protagonist according to the two perceptions are located on the same branch of the hyperbola. In fact, they result from two different breakdowns of its power relationship to the reference actor. The first interpretation of this remark is, after all, quite intuitive; there cannot be antinomic interpretations of the situation in which, for example, each of these actors would simultaneously consider themselves to be both dominant and dominated.

The representative points of view of the same antagonist are all the more distant as the divergence of points of view seems to be accentuated between these two actors, concerning their own perceptions of their bilateral relationship of forces.

Thus, when the representative points of a given protagonist do not belong to the same quadrant, there is cognitive dissonance between them and the reference actor. It is a divergence from their respective perceptions of the representatives of their power relationship. This divergence proves to be substantial when the representative points are located on non-adjacent quadrants. Paradoxically, these two actors can at the same time consider themselves to be "stronger" (with a simple dominance of stature) or "smarter" (with a simple dominance of posture) than their protagonist; hence the optical illusion that can lead to a "deaf ear dialogue", or even to an escalation of the tension between them. Finally, a relative cognitive consonance emerges when the protagonist's projections from both points of view are in the same quadrant. This consonance is made all the stronger the closer the points are brought together.

The analysis carried out on the matrix crossing the actors (in rows) and the objectives (in columns) is based on a singular approach. It led to the construction of a "hypergraph" of crossed utilities reflecting the tangle of power relations between actors. The same analysis can be transposed to a treatment of this dual-type matrix. From the matrix, this treatment makes it possible to construct a hypergraph of cross-preference relationships no longer between actors, but between objectives. It is possible to interpret this hypergraph in terms of posture and stature, that is, in terms of preponderance relationships between objectives.

7.1.4.4. *Characterization of push/pull force ratios*

For the reference actor, it is then important to carry out a comparative analysis of the posture/stature grids from the dual point of view of the reference actor and its antagonists (distribution of arrows within the graph), as they result from the push (singular) and pull (dual) approaches. This comparison makes it possible to determine to what extent and in which cases there is a concomitant relationship in the positioning of the reference actor with respect to their antagonists; and this according to both push and pull approaches.

Push/pull ratios call for a special interpretation. For a given variable or result, its push/pull ratio represents the ratio between the level reached as determined by the push or pull optics. This ratio reflects the degree of amplification that this variable undergoes when switching from pull to push optics. This justifies the naming of the push/pull ratio in terms of the switching coefficient of this variable.

On the basis of the previous definitions, the push/pull ratio of the strength or scope of a given player is therefore the multiplier coefficient of its strength between pull and push optics. It measures the leverage effect or the transition of the actor's force level, depending on whether it is considered as a receiver or a transmitter of interaction.

Taking these considerations into account, the push/pull ratio of the power relationships between players is established. This measures the "added value" coefficient of the same ratio between the pull and push optics.

This push/pull force ratio between two players is the product of the two push/pull ascending factors of stature and posture respectively. These two factors are, in other words, amplification coefficients for the tangible "stature" of the Goliath-type (quantity aspect) and the subtle David-type (price aspect) positioning, when switching from pull to push.

7.1.4.5. *The push/pull posture as analogous to the terms of implicit bargaining*

The push/pull posture is somewhat analogous to a well-known concept in the field of international trade: the terms of trade[3]. Indeed, posture factors are analogous to price indices and the pull and push optics refer to the notion of import and export. Thus, the push/pull posture ratio can be likened to a ratio between price indices, respectively, for exports and imports; which, for a country, corresponds to its own terms of trade, on the international market. Thus, the posture push/pull ascendance, as assessed via Astris, determines the indicator of a given player's "implicit terms of trade" with the reference player.

---

3 Ratio of the export price index to the import price index.

Another analogy can be drawn with the notion of consumer purchasing power. This indicates the ratio between the indices of increase in income linked to the performance of one's work on the one hand, and the increase in consumer prices linked to the purchase of products and services on the other. The importance of this indicator as a generic and collective support for wage negotiations is clear.

In this sense, the terms of the implicit bargaining of a given actor thus constitute a critical condition of acceptability – for the actor in question – of the bargaining underlying the exchange in which they participate.

We understand, in a completely intuitive way, that purchasing power expresses the maximum increase in consumption (in quantity) that an active person is able to make with the same income, without causing an imbalance in his or her budget. Like purchasing power, the implicit bargaining terms of a protagonist with respect to the reference actor correspond to the maximum pull effect of stature that the former is able to accept as a receiver of influences in this bilateral relationship, without degrading its push/pull balance of power. This means that the implicit terms of trade reflect the actor's capacity to allow themself to be carried along by upward currents and to absorb, if necessary, the shocks of their surrounding environment. We can thus detect the extent to which the dominant protagonist would be able to "let go", should they find themself confined in the role of a receptacle of influence; hence the notions of breathing space and potential resilience[4] associated with this indicator. Conversely, a protagonist with a favorable power balance, while having an unfavorable indicator of their bargaining terms, would find themself insidiously forced to spend a lot of energy "rowing against the tide" to maintain its dominance. Such a posture also suggests that this protagonist will run out of breath and thus eventually lose their dominant position.

7.2. Extending the analytical framework to a metasystem incorporating actors and objectives

This second part of the analysis concerns the extension of static analysis to the metasystem incorporating both actors and objectives. This is represented by an all-encompassing matrix comprising, in rows and columns, the same ordered set of actors and objectives, and integrating the subsystems of actors and objectives (or options), as well as their interdependencies.

4 "Resilience is the ability of individuals and systems (families, groups, and communities) to overcome adversity or risk. This ability changes over time; it is reinforced by protective factors in the individual or in the system and environment; it contributes to maintaining or improving health" (Mangham *et al.* 1995).

The purpose of the analysis is to decipher or decrypt, articulate and evaluate, in an intelligible and operational way, the predominance relationships of the network of competitors, based on all their interactions (actors/actors submatrix) and the configuration of their positions with respect to the different objectives (actors/objectives submatrix).

In this case, all events progress as if the metasystem were a hypergraph of interlinked power relations or utilities resulting from a two-layer sedimentation. These are the result of a complex articulation and recomposition between the cross-utilities calculated on the basis of the actor-actor and actor-objective submatrices discussed in section 7.1.1.4 on static partial analysis. This amounts to asking the following question in an interpretable and intelligible way: in the power relationship between two actors, how do their direct relationships and their relationships to objectives interfere, or even collide, and then aggregate and/or decouple?

7.2.1. *Globalizing the existence of each actor on the basis of their relations to the subsystem of factors and objectives*

7.2.1.1. *Visible pre-eminence over other actors, hidden via objectives*

From this perspective, Astris allows the relations of pre-eminence of an actor with the other protagonists to be evaluated around two new and complementary axes:

– the direct and visible pre-eminence exercised through the multilateral interactions of the actors;

– the induced and hidden pre-eminence from which the actor benefits through the consideration of the corpus of objectives.

7.2.1.2. *Visible predominance over objectives and hidden via other actors*

Astris makes it possible to evaluate the relationships of preponderance of a particular actor on each of the objectives around two similar axes:

– the visible preponderance that this competitor exerts through its direct hold on the corpus of objectives;

– the hidden preponderance it enjoys through its influence on the various protagonists.

The conceptual graph presented in Figure 7.7 illustrates the breakdown of pre-eminence or preponderance according to the nature (visible/direct or hidden/induced) and object of their effects (actors or objectives).

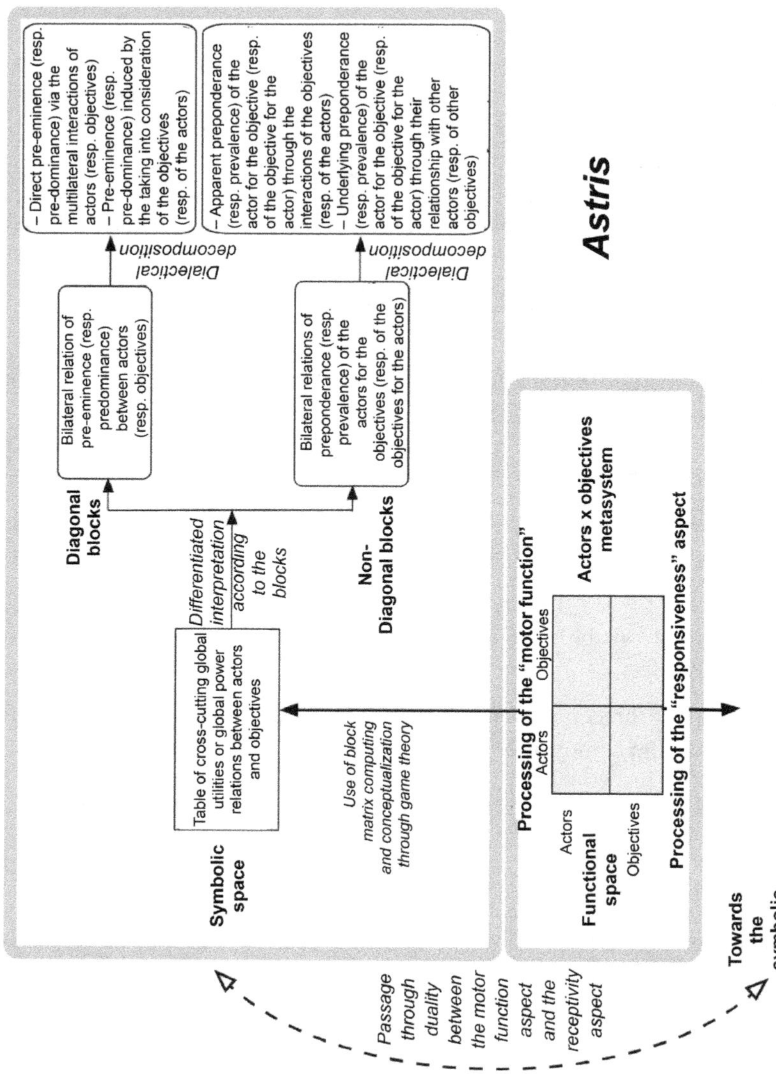

Figure 7.7. *Overall static analysis: conceptual diagram*

7.2.2. *Aggregating the existence of each objective based on its relationship to the two subsystems of objectives and actors: visible and/or hidden preponderance and prevalence*

Breakdowns of the same type with two different presentations, visible and hidden, are also provided with regard to the interactions:

– between objectives, which is described as a "preponderance" visible (and direct) over the other objectives and hidden (and indirect) via the actors;

– of objectives with the actors, known as "prevalence", visible (direct) on the actors and invisible (indirect) via the objectives.

Moreover, since this is a closed metasystem associated with a square matrix, it is possible to switch almost automatically:

– from the *singular* (or *primal*) approach, representing the motor aspect (push optics), which is the subject of this and the previous section (7.2.1);

– to the transposed (or *dual*) approach, reflecting the receptive aspect (pull optics).

7.2.3. *Extending the relational existence of each actor by adding successive dimensions*

The same all-encompassing analysis can be carried out, for example, within the framework of a system made up exclusively of actors. This system would thus integrate two subsystems based on the proximity of its constituent entities, so that it can be considered as a metasystem of actors, itself nested within the first metasystem encompassing the actors and objectives. These actors could also be classified into two subsets based on whether or not they belong to a group, a community, or even the same professional body or group.

7.2.3.1. *From the most tangible and closest dimensions...*

In this way, the evaluation of the relations of pre-eminence of a given actor over any other actor affiliated to the same group can be broken down into two complementary effects:

– a visible pre-eminence effect that the actor exerts through its interactions with the actors of its own group;

– a hidden pre-eminence effect from which the actor benefits through its interactions with actors who are not members of this group.

This interpretation key generates a new conceptualization of the "power" of a given acting entity; in this case, an actor. The indicator of this power is provided by the diagonal element, representing the entity in question in the global cross-utility matrix. This element is what we call "self-satisfaction" or overall "sufficiency" in the situation of this actor, with regard to all the information taken into account.

The notion of the "sufficiency" of each actor can be interpreted in a polysemic way. This represents, in particular:

– the intrinsic level of satisfaction that this actor acquires in the exchange;

– the fair degree of recognition to which it could be entitled, because of its participation in the acting;

– the implicit relational force that it would manage to extract from this exchange.

This sufficiency is organized and configured by sedimentation in successive layers, ranging, for this actor, from the most tangible and endogenous dimensions to more intangible and exogenous dimensions. The second of these sometimes proves to be just as important.

The first dimension is made up of the material elements characterizing the actor-entity in an intrinsic way (its weight, its means, its solidity), that is, in the numerical consideration of the elements appearing in the diagonal of the initial matrix of interaction of the protagonists.

The second and third dimensions concern the aspect of sufficiency attributable to the interactions (influences, exchanges) that the actor entity maintains, both with other actors of its reference group or close circle, and through entities outside this group.

This aspect is of particular importance to the extent that it brings to the fore the concept of "emptiness", according to which the existence of an actor-entity is more a product of the intricate web of its interactions with other entities than from any specific substance associated with it. It is relational capital that is favored to the detriment of a "reified" existence.

7.2.3.2. *... to the most intangible and remote dimensions*

The fourth and final dimension no longer involves networks of exchange and relationships, but rather the hypergraph induced by the possible sharing of concerns or objectives, towards which these actor-entities or – by extension – their reference groups have the potential to mobilize. At this stage, it is a dimension that involves more impalpable and intangible aspects, and which contributes to the emergence of virtual circles, affinities, or adherence to a set of ambitions or values.

The question concerns the way in which the "coopetition" (cooperation and competition) between acting entities or actors takes place, which is to say, the dialectical connection through:

– an antagonism inherent to the competition of these on similar objectives;

– a synergy, or even a de facto cooperation, favoring the creation of favorable conditions for the predominance of such objectives.

In short, the power of an entity (firm, organization) or an actor can be conceptualized in the form of a stratification of identity components, ranging from the existence of the actor-entity itself to the vacuity effects that help to shape it. These effects can come from exchanges instantiated by networks of tangible relationships, as well as from the sharing of interests or affinities within virtual communities. This is why it would seem appropriate to introduce the concept of the "eco-existence" of this entity. This concept is defined in an intrinsic or inherent way to the entity concerned, as well as in its relation to other entities or to its context.

7.2.4. *Measuring the overall coherence and the capacity for relational convergence of the metasystem*

First, it is important to indicate the coherence of the interactions of the elements of a subsystem, such as that of the actors and/or objectives, which corresponds to the notion of internal coherence.

7.2.4.1. *Estimation of the internal coherence or relational consistency of the subsystems*

The construction of the internal coherence indicator measures the harmony that exists between this subsystem, considered as a whole, and its different elements. This indicator measures the "internal utility" of this subsystem, with regard to its various integral parts. The evolution of the value of this indicator over time constitutes an appreciation of the propensity of the system in question to converge around a possible compromise (such as the players in the game, institutions within the framework of international negotiations, stated objectives, regulations in force, etc.). This is an issue that remains relatively poorly understood at the conceptual and practical levels. Furthermore, as with the question of the precautionary principle, it has been the subject of numerous controversies and remains at the center of concerns expressed in various studies carried out within the framework of international institutions or organizations.

7.2.4.2. Indication of the overall coherence of the actor-objective metasystem

In the second step, we estimate the global multi-actor coherence. This results from the evaluation of the consistency of the interactions of the actors. These interactions are considered to be immersed in the metasystem of actors and objectives. The decision-maker can also acquire a decomposition of the global multiactor coherency indicator in two ways:

– the indicator of intrinsic multi-actor consistency resulting from the direct interactions of the actors;

– the multi-actor extrinsic concordance indicator induced by the positions of the actors with respect to the objectives.

Similar results concerning multipurpose coherence are also provided.

All such results are aggregated to measure the overall consistency of the multi-actor and multi-objective interactions. They concern the overall consistency that is no longer of a particular subsystem – such as that of the actors immersed in the metasystem of actors and objectives – but of the metasystem as such. This new information makes it possible to understand how and why the global coherence of the subsystems with complex interactions can be affected by the relational dysfunctions that insinuate themselves between them.

7.3. Interpreting the dialectic of transformation of the interplay between actors

The static analysis performed by Astris is based on the observation of a single period. When the analysis is done on several periods – extending, for example, over 15 years – the tool allows the same graphs to be obtained regardless of the periods recorded. It makes it possible to create a single graph visually displaying the representative sequence of the evolution of the positioning (representative points) of the same actor over all the observed periods. However, this is not a truly dynamic analysis, which is why this development is necessary.

According to Perroux (1974), "major periods of development are characterized by the change in the course of the evolution of the relative weight of structured sets. The analytical representation of this evolution is a matrix (or a network) that is formed over time. This orientation reflects the changes in structure that can be predicted *a priori* to be accompanied by changes in relative power".

The exchange is characterized by the ambivalent nature of the actors' behavior and the dialectical nature of their "cooperative" interrelationships. This dialectic generates a certain level of energy for change which, through a succession of temporary balances and irremediable imbalances, in turn generates the dynamics of the *composite exchange.*

7.3.1. *Diagnosing dynamic power accesses*

This last part of the analysis is done on:

– the deformation of one of the subsystems of actors or objectives (partial dynamic analysis);

– the deformation of the metasystem incorporating both actors and objectives (global dynamic analysis).

The idea is to free ourselves from the limitations involved with the succession of snapshot photographs in order to make a movie sequence aimed at better detecting the underlying progressions.

7.3.1.1. *The notion of dynamic override or power access*

The analysis allows us to assess the ability of each actor to surpass themself, by coming to identify an additional weight following the exchange with the other actors. For each actor involved in the exchange, this surpassing, called "dynamic overpower" and evaluated quantitatively, is obtained by the difference between two terms:

– the effective change in the weight of the actor in the exchanges in which it participates;

– the potential change in weight from which this actor could benefit due to the mechanistic repercussion – through the tangle of its power relations – of the modification of the weights of the different protagonists.

In other words, dynamic overpowering measures the extra weight or "access of power" that an actor manages to draw from the exchange, compared to that which would result mechanically from their relations to others or an intersubjective relationship.

This dynamic overpowering is therefore not limited to the simple evolution of the weight of the actors in the exchange: this also has the potential to create optical illusions. Indeed, it measures the net capacity to exceed (in terms of weight) demonstrated by the reference actor in the framework of the composite exchange. This "power access" can be compared to the change in energy that the actor demonstrates in their potential for creating "increased power". Thus, the dynamics of the composite exchange are accompanied, from one year to the next, by transfers of energy between actors. These transfers would then be inclined to rebalance, so that the power access of the exchange system, considered in its entirety, tends to remain stable over time. This last observation echoes the hypothesis put forward by Perroux (1981), which holds that the dynamic of composite exchange could refer to a thermodynamic-like paradigm. Thus, a sufficiently encompassing and compact network of actors, due to the entanglement and density of its interconnections, could be intuitively considered as a quasi-closed system. Under these circumstances, the extent of the effervescence of the "cooperative" transactions that it conceals would remain, with reference to the notion of entropy, relatively constant.

7.3.1.2. *Breakdown of effects of confrontation (ego) and conciliation (eco)*

Moreover, this highly dynamic access to power can be broken down into two new effects, both complementary and contradictory, which reflect the very essence of the dialectic of composite exchange:

– An effect of differentiation or confrontation, consisting of the actor asserting itself and distinguishing itself through its own initiatives, sometimes considered undesirable by the protagonists. This first effect reflects a certain aggressiveness on the part of the reference actor, based on a rather egocentric type of behavior.

– An integrating or reconciling effect aimed at synchronizing and harmonizing with others, to avoid certain dissonant effects that are detrimental to the cohesion of the system. This second effect, of an ecological or ecosystem-level type, is seen from a more empathetic perspective, where the actor of reference is more inclined to make concessions, taking into account the plurality of appreciations made by the protagonists on its own behavior.

The dialectical nature of the connection between these two effects, which are notably ambivalent and two-sided, should be emphasized.

The conceptual graph presented in Figure 7.8 illustrates this breakdown of the access to power.

In terms of processes, the evolution of exchanges is thus characterized by a succession of imbalances, punctuated by provisional balance points.

Thus, by seeking to synchronize with others, sometimes to the detriment of its own initiatives, the reference actor deprives themself of real potential for overcoming. On the other hand, an aggressive rise in power on their part which is seen as too arrogant could trigger, after a period of time or beyond a certain threshold, an outcry from the other protagonists. This defense could potentially go from a simple movement of irritation or protest to a more marked rejection by the other protagonists. Excessive aggression by the reference actor would expose them to a restriction of their own maneuvering space, which would potentially call into question their previous advances. These two dialectical effects cannot be reduced to a simple arithmetic artifice, as they are considered both consubstantial with the "access to power" and constitutive of the dynamics of composite or cooperative exchange.

Thus, the change in each actor's access to power and its dialectical breakdown in terms of egocentric and ecosystemic effects follows an implicit rule, which could be summarized in the following phrases: "no sustainable ego, no eco-sustainability!" or "ego without eco means a fiasco!"

Through different applications of this tool-based approach, we have observed that dynamic efficiency, as well as the two resulting singularization/differentiation and synchronization/integration effects, generally evolve more regularly than the raw data relating to the exchange flows between actors, which are rather fluctuating or erratic. Moreover, the evolution curves of our indicators show more marked and lasting reversals of trend that seem to reflect profound, even structuring changes, both in the nature and in the "terms" of this exchange. This analysis makes it possible to validate the interconnection – one that is admittedly complex, though productive – between the observation of exchange flows at the operating level and the symbolic expression of higher order (metalevel) power relations[5] and strength relationships.

Thus, the intelligence of evolving situations deliberately and metaphorically consists of giving priority to detecting base-level waves – or even tidal movements – over superficial observation of the swells. For an analyst to focus on this second objective of observation would be, in this case, delusional.

5 According to François Perroux (1973), "An exchange is first of all inseparable from the concept of power, which appears – beyond the simple accounting of prices and quantities – in the form of transfers of influence between suppliers, considered as actors in their own right. This transfer takes place through asymmetrical interrelationships, the intensity of which stems from judiciously defined cross-utility functions".

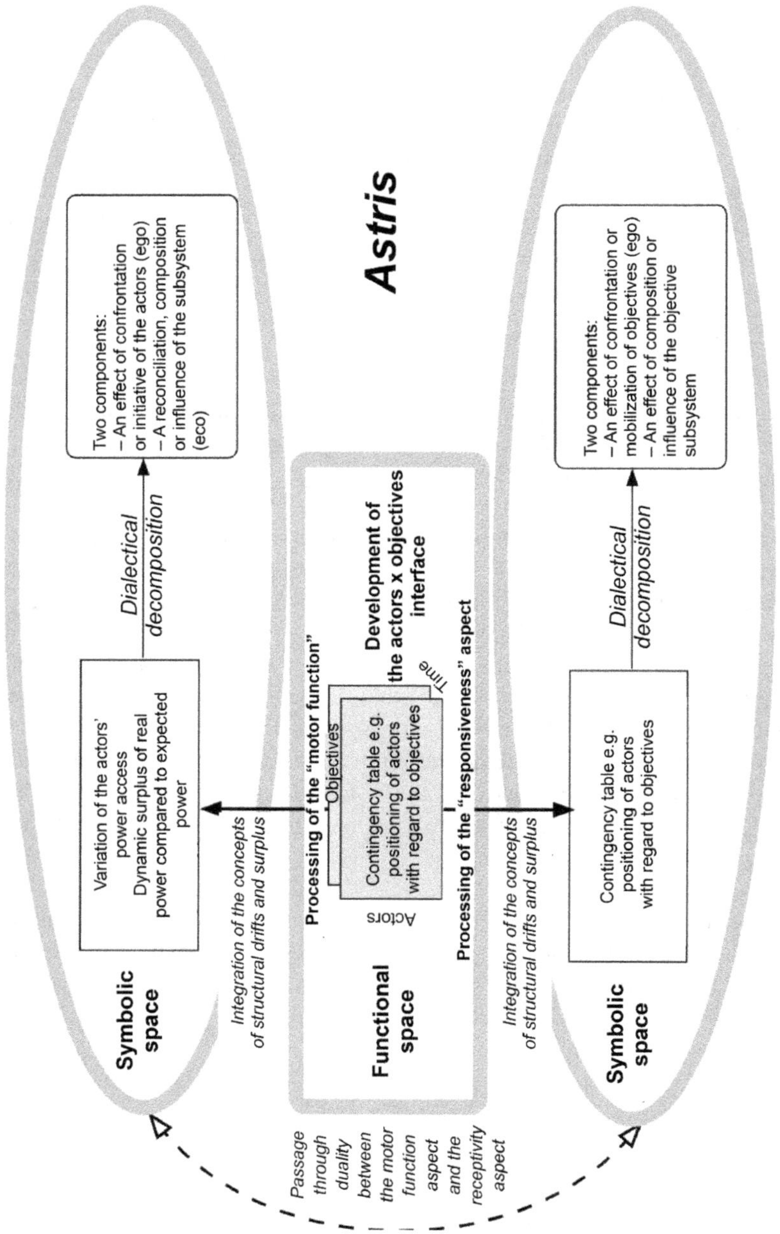

Figure 7.8. *Partial dynamic analysis: conceptual framework*

7.3.2. *Interpreting the dialectic of power access regimes*

The results of the dynamic analysis also offer a rare opportunity to interpret the dialectic of power access regimes. This is carried out following two complementary visual approaches of periodic classification of dynamic overpasses and multi-year induction of the presumed type of development.

7.3.2.1. *Periodic classification of dynamic overrides*

This approach allows for the creation of a heuristic grid of periodic classification of its different actors for each period and for the dynamic of the various different actors. This grid, for each period, is the projection of the different actor-points in the ego/eco planes, with "egocentric confrontation" as the y-axis and "ecosystem conciliation" as the x-axis.

This graph, shown in Figure 7.9, is divided into four main positioning indicator quadrants. These correspond to the different quarter-planes delimited by the y-axis (ego) and x-axis (eco), respectively labelled as irresistible, tolerable, beyond repair, and unforgivable.

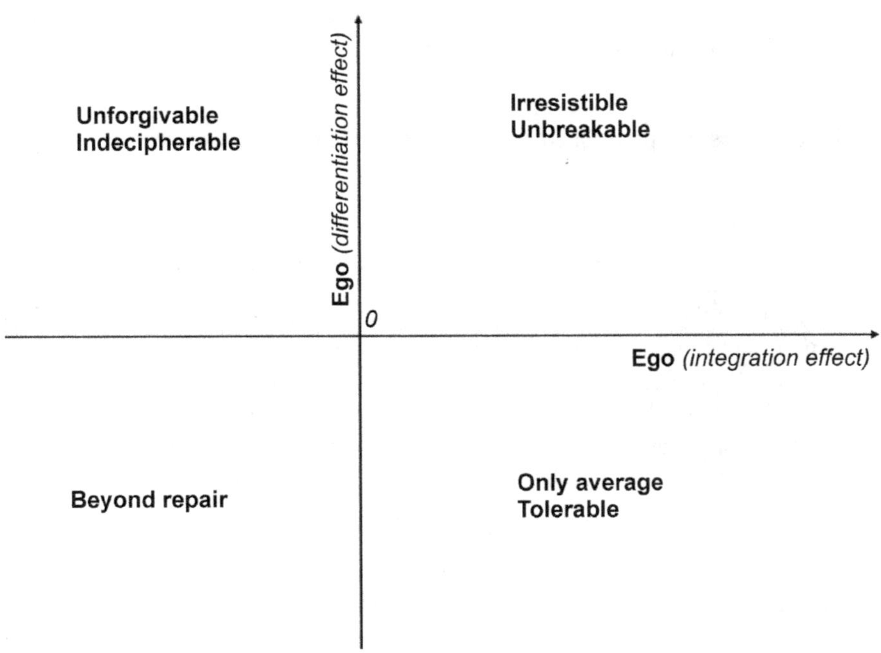

Figure 7.9. *Ego/eco graph for dialectical interpretation of "power accesses": grid for periodic qualification of power overruns (four subsectors)*

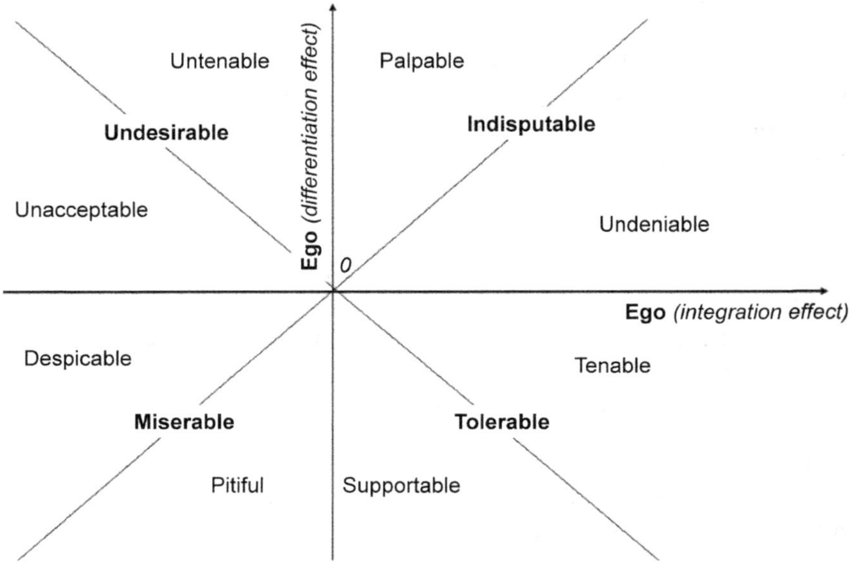

Figure 7.10. *Ego/eco graph for dialectical interpretation of "power accesses": graph of periodic qualification of power overruns (12 subsectors)*

These sectors are partitioned at a finer level into a dozen or so sub sectors of the ego/eco plane characteristic of the positioning of the actor points, as shown in Figure 7.10.

7.3.2.2. *Multiperiodic presumption of the type of induced development*

These grids allow us to visually retrace and interpret the succession of the positioning, and thus the trajectory, of different points during all the periods. The trajectory of the representative point of a given actor in the ego/eco plane often describes a "Brownian" type of movement, crossing a particular succession of sectors and/or sub-sectors over time. For purely illustrative purposes, Figure 7.11 shows the point-year trajectory of the group of developing countries (DCs) over the period of 1972–1981.

On the other hand, when the movement of this actor-point proves to be significantly circumscribed within a particular subsector of the ego/eco plane during most of the observed periods, it then becomes possible to anticipate its *presumed* type of *development*; that is, to discern the mode of relational development likely to be induced by the quasipersistence of its over-powering regime.

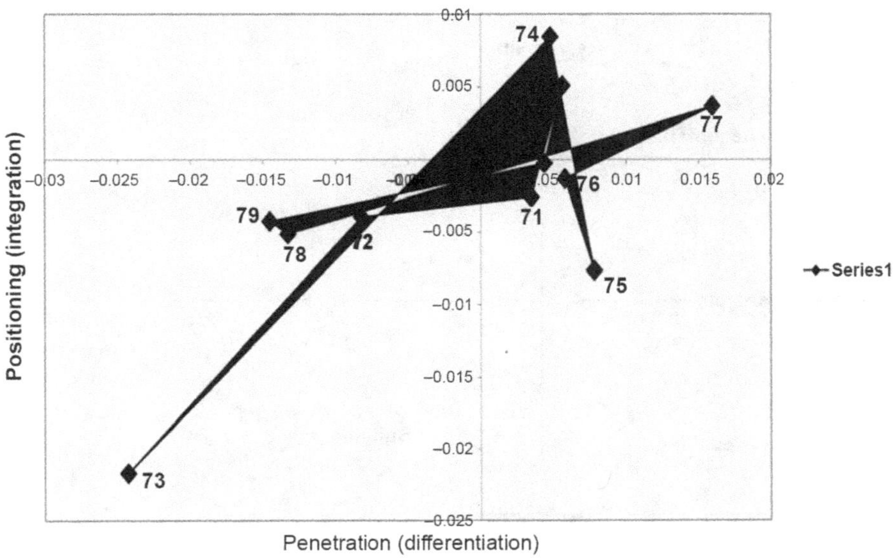

Figure 7.11. *Ego/eco graph showing gains in power by developing countries during the period that includes the two great oil crises of 1973 and 1979*

Beyond the normative and widely-known concept of sustainable development, other development methods or situations – responsible, fair, reasonable, appropriate, deplorable, condemnable, or execrable – can be identified through the second heuristic grid presented in Figure 7.12. These development methods and situations are more nuanced and equally significant. They are also clearly identifiable through our heuristic grids. They can therefore be enriched, refined, or revisited through the proven practice of the dynamic approach to the access to power.

In this respect, we found that the current emphasis on the concept of sustainable development was likely to encompass, absorb, and even eclipse other development modalities. Thus it is necessary to diagnose these, in order to be able to better understand them and to try to influence them in the desired direction, in order to cause them to progress, through the proven practice of the dynamic approach of power accesses.

In conclusion, the advanced method, propped up by Astris, helps to formulate detailed recommendations to stimulate synergies and prevent antagonism by defusing possible conflicts. Thus, thanks to its dynamic analysis functionality, the essential contribution consists in detecting early warnings to manage a real dynamic of trade-offs.

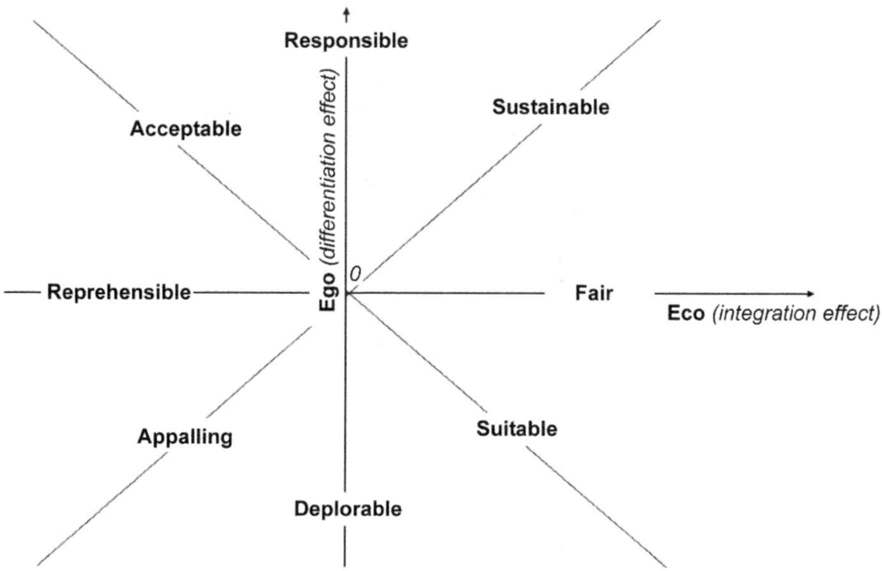

Figure 7.12. *Multiperiod presumption grid of developmental type, according to the permanence of overpowering regimes*

This approach helps to identify and decipher the rough edges of the situation and, depending on the what the user wishes to accomplish, can help:

– address discord, dissonances, and divergences in an attempt to reabsorb them and "get back to the drawing board";

– channel them to take advantage of them, by inserting them into a strategy in which transition and imbalance are the main driving force behind the dynamics.

In light of the phenomena detected in this way, analysts, business intelligence specialists, and decision-makers are able to avoid optical illusions and better understand and support the real praxeological transformations, in order to bring about – if necessary, and at the right time – a change in the desired direction. In this respect, our approach is a precious instrument of regulation[6]. It helps to clarify and evaluate the contributions of the different actors to cooperative exchange. It also

6 The notion of regulation refers to the endogenous dynamics and the mechanisms of the self-maintaining nature of the behaviors of a set of actors, which tends to integrate such behaviors.

makes it possible to formulate circumstantial recommendations to stimulate synergies and prevent antagonisms by defusing possible conflicts.

Thus, thanks to its dynamic analysis functionality, this approach allows for early warnings to be detected in order to stimulate, attempt to regulate, and drive a green spiral of compromise, which is likely to be ultimately conducive or even beneficial to most players/actors.

8

Breathing Lasting Vitality into the Long Term

The issue that arises at this stage of the strategic engineering approach is to know to what extent and how this instrument-based approach, beyond the recommendations it offers before decisions are made, can help monitor and evaluate the strategic direction of the intervention once it has been implemented. How can it help to progress "on the way" towards a satisfactory outcome to a crisis situation we might face?

This is a fruitful perspective for the dynamic use of the PraxéoGraphis concept (see Box 8.1), allowing us to move from an approach consisting of interactions between components (a systemic model) to an approach in terms of evaluative properties, leading to a true dynamic coupling between the "intervention" and "strategic landscape" subsystems.

PraxéoGraphis is an experimental platform for strategic engineering based on the original concept[1] presented in the previous chapters. Its goal is to help discern, arbitrate, and intervene in complex situations, especially under adversity or in a perplexing environment. The use of PraxéoGraphis helps to regenerate, integrate, and redirect the "strategic flows" of the company or organization, allowing it to rebound under all conditions and to succeed in the long term.

PraxéoGraphis is based on systemic analyses at multiple levels or scales (following the notion of "systems of systems"), through the concept of nested frames. It allows for the representation of the "strategic landscape" as a framework for systemic analysis of the situation and contextualization of strategic action. This landscape is thus conceived as a system that can be broken down into three macrocomponents: the strategic referential, including the vision, mission, and vocation of the organization or company, which is the mastermind of the reflection; the exploratory scenarios or general environment scenarios,

1 Developed by Jacques Arcade.

which may occur over the period studied; the context or the theater of intervention, incorporating the "interplay of actors" in particular.

Each of these components is in turn broken down to different scales according to the level of precision required by the circumstances.

In particular, PraxéoGraphis offers the possibility of visualizing the strategic landscape according to two complementary optics: a hierarchical breakdown of the components, in tree form; and a mapping of their interactions at different scales, in matrix form. This integrated tool allows the construction of strategic intervention options, as well as the structuring of them according to the same logic, in the form of a subsystem of nested components.

The comparative evaluation of the different interventions under consideration is carried out through the analysis of their capacity of immersion in the strategic landscape. This results from an "ecosystemic" assessment of the interactions of the different interventions with each of the components of the strategic landscape. This assessment is carried out dialectically from two points of view: the influence of the strategic landscape on the interventions and the impact of the interventions on the strategic landscape.

The PraxéoGraphis platform is conceived as a reception structure integrating three specific instruments corresponding to particular fields of capabilities, such as Morpholudis, Simoptis, and Astris, which are examined in the boxes provided in Chapters 2, 6, and 7, respectively.

An initial IT integration of this original concept, Simulia-SDS (Simulation – Strategic Decision Support), was carried out on behalf of the Ministry of Defense in a partnership between Atos Origin and Proaxis.

Box 8.1. *PraxéoGraphis: an integrated strategic engineering toolbox*

This approach, inspired by the life sciences and similar to the biological model, is phenomenological in nature; it takes into account the mechanisms by which changes in the strategic landscape tend to disrupt the relevance and adequacy of the solution that is implemented in order to provoke readjustments that are more or less substantial. Conversely, the intervention fertilizes the context by infusing it with the alignments of a local and "silent" transformation. By deepening and gradually changing the viewpoint of the decision-maker/strategist on the evolving "situation–solution" pairing, this approach proves to be a generator of cognition and organizational learning. In particular, it offers the possibility of gaining access to the politico-strategic viability of actions.

Conceptually, this dialectic represents a spiral progression, following a vortex-like movement. The entanglement of such movements over time generates a dynamic of mutual enrichment of the evaluative properties, as well as of the behavior of the system, which translates into an increase in the content of the evaluation in the sense of greater subtlety. This constitutes a true meta-evaluation (of order 1). Its purpose is less to evaluate the overall performance of the solution at different moments in its execution (through its finality, its pluralism with regard to the future, and its pragmatism) than the degree of integration, over time, of the monitoring-evaluation and adjustment processes at work for each of these criteria.

For example, the emerging evaluation of "directivity" is superimposed onto and follows the more static evaluation of finality. Similarly, the assessments of "insight" and "agility" incorporate and transcend those of pluralism and pragmatism, respectively. Thanks to such a conceptual elevation, we can now move from a pre-decisional appreciation of performance (which is admittedly broad, but based on a somewhat "substantive" rationality), to a more effective assessment, based on a "procedural" rationality of the "viability" of action (as a continuous dynamic of the achievement of this performance).

The extension of the same process of elevation beyond monitoring and evaluation allows us to conceive of a second metalevel (of order 2), in which the notions of prosperity, durability, and liveliness would be implemented as if by flocculation, as emerging properties of the intervention mechanism; these would then form the new foundation of the evaluation, placed above it.

The ultimate synthesis of this higher-order assessment would revolve around the more essential notion of "vitality". Such a notion would seem more likely to promote, under all circumstances, the achievement of the goals and the consecration of the intervention mechanism, potentially carried out "from above" and to prove to be sustainable, by means of a path out of the crisis.

This perspective of extension beyond the current perimeter of the Praxéo-Graphis instrument, inspired by the method of strategic engineering as a whole, constitutes a true "guide for enlightened action" at the politico-strategic level.

8.1. A gradual movement towards a good start

A fundamentally dialectical relationship can be established between the macrocriteria used in evaluation. From this perspective, it is important to review the skill concentrations required to guide action in an informed manner.

8.1.1. *Mobilizing a triad of key skills to guide action*

Our strategic engineering approach brings together the three main distinct disciplines of foresight, strategy, and monitoring. Each of these includes a wide variety of proven concepts, methods, tools, and recognized practices.

Strategic engineering focuses on three poles or areas of expertise: contextual intelligence, strategic engineering, and decision engineering.

8.1.1.1. *Contextual intelligence*

Contextual intelligence consists of the original know-how by virtue of which the actor (the company, the administration, the local authority) is able to carry out a systemic observation of the different contexts (economic, technological, organizational, sociopolitical, regulatory, etc.), situations, and circumstances in which the action emerges or is exercised. In particular, this approach leads to a synoptic representation of the concrete system of action in the form of integrated contextual frameworks, thus facilitating understanding. Contextual intelligence is an integral part of the art and the mechanisms used for conceiving, implementing and carrying out any exploration and anticipation of the strategic landscape. It leads to exploratory scenarios describing the possible futures that the company may face.

8.1.1.2. *Strategic engineering*

Strategic genius includes, in addition to the art and manner, the talent and intuition needed to develop strategic inspiration and then deploy the action. The essential result of this approach concerns the strategic design or vision, the major orientations, the core business of the actor-company, as well as the design of projects allowing for them to be implemented. This is the "strategic reference frame" that characterizes the finalization and foundation of any strategic action.

These skills, whose specific methodologies and tools are beyond the scope of this book, can indeed be qualified as fundamental (according to Hamel and Prahalad (1999)), in that they are:

– generators of essential functionalities to guarantee the deployment of efficient actions;

– distinctive, compared to the types of skills most experts and specialists in these fields possess;

– able to be used as pivots (i.e., that can be combined with each other and with other skills) to open up new spaces for transdisciplinary investigation and advice.

8.1.1.3. *Decision engineering*

Regardless of the genesis, nature, or form of the company's actions or structuring interventions, they must prove to be efficient overall. From this perspective, the evaluation of the performance will be based on an original combination of macro criteria derived from the three key competencies described above, in terms of inspiration, prospective exploration, and systemic observation of the context of the action.

Achieving a sustainable performance depends on the company's ability to combine contextual intelligence and strategic engineering, and above all to make the right decisions, to make judicious choices and subtle trade-offs in a context of uncertainty that is perplexing for the decision-maker. Thus, strategic engineering is part and parcel of a dynamic of an "enlightened guidance" of action, centered around a proactive mindset. This last property results from the articulation and overcoming of reactivity and pre-activity.

8.1.1.4. *Guidance as a key ability*

Guidance is a fundamental competence of the actor-company that, through the activation of directivity, insight, and agility, is able to transcend them, ensuring the viability of the strategic action and ultimately contributing to the vitality of the company.

If the guidance is broad-reaching and is at a higher level of complexity, it also constitutes a relatively new, solid, and beneficial way to approach the question of extreme difficulty:

– to grasp, in an evolutionary way, the real drivers of strategic action;

– to approach, catalyze, and support the incomprehensible alchemy of sustainable development and corporate vitality in the best possible way.

8.1.2. *Bringing together the three dimensions of performance evaluation: finality, pluralism with regard to the future, and pragmatism*

In order to achieve a satisfactory level of performance, as developed in Chapter 4, the design and evaluation of any strategic intervention must refer to a triptych of broad dimensions or macrocriteria that correspond respectively to strategic engineering, prospective engineering, and contextual intelligence, namely: finality, pluralism with regard to the future, and pragmatism.

Broad dimensions or macrocriteria	Criteria in dialectical interaction	Specification of the criteria for apprehending strategic approaches
Finality **Relevance to vision and mission**	Voluntarism Conservatism	Well-founded in relation to the problem, concordance with the major objectives Compliance with the mission, respect for the company's vocation
Pluralism facing the future **Adequacy for future environments**	Idealism Realism	Proactivity, ability to shape or influence the environment favorably Preactivity, validity, and positioning in relation to the exploratory scenarios
Pragmatism **Congruence with the environment and the context of intervention**	Utilitarianism Opportunism	Utility, functionality, convenience for the environment (actors) and context Admissibility, acceptability by the environment and context (economic, regulatory, etc.)

Table 8.1. *Summary table of macrocriteria*

These three macrocriteria are both complementary and contradictory, thus the subtlety of their dosage – as expressed via their respective weighting – in the overall assessment of performance. These major dimensions (or macrocriteria) result from an essentially dialectical articulation between two subdimensions. These can, in turn, be broken down into sets of operational criteria.

8.1.2.1. *Finality or relevance to the main orientations*

The first major dimension concerns the macrocriterion of finality. This takes into account the relevance of the strategic initiative in relation to the company's vision, vocation. or mission. This macrocriterion is underpinned by the dialectic composition of two criteria: voluntarism and conservatism.

Voluntarism is based on the merits of the strategic initiative in relation to the vision. It reflects the qualification and the "rightness" of the strategic initiative considered as an asserted response or as a solution to a problem, based on its concordance with the primary orientations and *ad hoc* objectives.

Conservatism can be observed through the degree to which the line of action conforms to generic requirements that may fall within the mission of the organization in charge of the strategy, its vocation, or its culture. These requirements sometimes take the form of consensual imperatives concerning budgetary, economic, social, or other such aspects.

These two criteria are complementary insofar as it is necessary, in order to be successful, to know how to combine the will to change and the "installation" within a certain continuity. Moreover, they tend to be in opposition. Indeed, by striving to respond effectively to an important but singular problem, the company can insidiously divert itself from its original mission.

8.1.2.2. *Pluralism in the face of the future or adequacy to future environments*

The second major dimension concerns the pluralism of the action intended to be taken for the future. The plurality of profiles of hypotheses constituted by the scenarios reflects the polymorphic nature of the future. It is therefore appropriate for the firm to adopt, through its strategic initiatives, a stance that integrates this reality. In this optic, the macrocriterion of pluralism should make it possible to assess the appropriateness of the line of action for future environments. This is a question of putting it into perspective, with regard to the different exploratory scenarios developed in the framework of a preliminary prospective study. This dimension covers two aspects which are also in dialectical interaction: idealism and realism.

Idealism, which more closely corresponds to the area of proactivity, translates the propensity of the envisaged action to inflect and shape the future environment by influencing the advent or the contours of this or that scenario.

Realism is based on long-term positioning and the validity of the action in relation to exploratory scenarios; it is more within the realm of pre-activity.

8.1.2.3. *Pragmatism or congruence with the context of intervention*

The third dimension involves the pragmatism and the operationality of the strategic initiative. It is based on the congruence of the course of action with its context of intervention; that is, on its ability to integrate effectively with its environment and the actors involved, the situation (economic, social, etc.), and even the particular circumstances of the action.

Pragmatism, which focuses attention on the immediate environment (proximity of content and close horizon), also revolves around two points of view or criteria: utilitarianism and opportunism.

Utilitarianism reflects the scope of the action, that is, its convenience and functionality (its utility) for the context of intervention and the actors.

Opportunism is based on the admissibility of this action by the context and on the acceptability of this action by the stakeholders or actors of this context.

Each of these two criteria will in turn be interpreted according to the different components (technological, economic, ecological, sociopolitical, regulatory, and institutional) resulting from the systemic representation of this context.

The three macrocriteria form a triad. Indeed, these criteria are distinct, but complementary, even contradictory. They are articulated to allow the evaluation of the overall performance of the courses of action, without obscuring their differences. The balance between these three macrocriteria, as well as the translation of this balance into the form of decisive arbitration, evolves according to the nature of the actions selected and the preliminary results of their testing.

8.1.3. *Driving and animating wise actions: a conceptual insight*

Once the alternative action or intervention (simple or composite) has been selected, the test of this action must be judiciously guided. This process should focus on the continuous systematization of the preliminary evaluation of the overall performance, based on the following:

– the understanding and continuous interpretation of the disturbances generated by the action in its situational context – most often complex – which it will be confronted with;

– the strategic questions that this action raises and the detailed review of its application methods, its expected results, and even its foundation or purpose.

8.1.3.1. *The logic of rebounding and the self-deployment of the action*

Whatever the quality of project formulation (i.e. its consistency and relevance), it is in evolution, both because of changes in the context and the conditions of its implementation, and because of its design (its chain of objectives, target levels, planned actions, etc.). Although such revisions sometimes seem incongruous in the light of a rational approach, they remain essential to the sustainability and success of the intervention. Indeed, the intervention must be considered as a living system whose survival and development will depend on its ability to cope with new demands. This ability could result, in particular, from the gradual adaptation to obstacles and the concomitant revision of ambitions. It is therefore important to deconstruct the main mechanisms governing the dynamics of the rebound and self-deployment of intervention projects. Finally, some lessons will be drawn from the heuristic nature of strategic action and the illusory virtue of overly normative and rational approaches.

8.1.3.2. *Strategy as a multifaceted concept*

On the basis of a predetermined objective and assuming a synoptic knowledge of the terrain, the trajectory taken by the intervention can be compared to an initial understanding of the concept of strategy. However, this trajectory is likely to deviate significantly from the global orientation of the intervention project and becomes all the more winding or chaotic as the path meets obstacles. The lack of information and the contingent nature of the environment lead us to admit that this path is being built gradually, according to events and even obstacles. Thus, the process adopted will lead to a destination (a "quasi-objective") which, in the end, will prove to be singularly different from the one initially envisaged.

In this logic of mutual fine or medium adjustments, the concept of strategy can no longer be limited to the recommendation of the trajectory in the *strict sense*. Therefore, strategy appears to be inseparable from the concrete process of clearing the path. Strategy corresponds to a corpus of guidelines or rules of conduct of an operational nature, aimed at ensuring the viability of the project-intervention.

In order to be effectively pragmatic, these rules will be structured around the key success factors. Therefore, the ability to continuously link these critical benchmarks with *de facto* evolving objectives is essential: it is the ability of "strategic tensioning" that emulates the organization's "driving force".

8.1.3.3. *The emergence of related actions and the drift of objectives*

The type of management described is based on the achievement of the project's primary objective, which is to solve its founding problem. However, this pattern implies a certain constancy of purpose. This approach can therefore only be considered plausible insofar as it is within the limits of a short period of time, corresponding to a circumscribed section of the intervention project's trajectory.

The emergence of a serious problem is likely to jeopardize the viability and therefore the sustainability of the intervention. In such situations, most of the concerns tend to focus on resolving the crisis of the moment; this leads us to consider certain complementary interventions of an *adventitious* nature, whose aim and content may therefore be distinct from those of the previous project. This leads to shifts in the trajectory and insidious drifts of objectives, which turn out to be complex and unpredictable. However, in this branching process, the methodology for designing and implementing complementary actions is similar to that of the initial project. It constitutes a functional invariant, since it is repeated each time the resolution of a problem is approached.

Seen from this angle, the distinction between ends and means deserves to be nuanced. Indeed, their mutual adjustments, when the critical points prove to be crucial and any modification on the margins becomes inoperative, lead to a salutary and radical redeployment towards other ends; this gives rise to the phenomena of a change of course or a bifurcation of the global evolutionary trajectory of the project-intervention. It is important to note that such a process of the creation of the trajectory of action and simultaneous generation of interventions develops through a "wreath effect", like that of a firework or a fragment bomb. In order to pursue and exploit this metaphor, it is important to observe that in the case of such a wreath effect:

– several offshoots may appear at different times, but evolve in distinct sections;

– these flashes are relatively ephemeral and, as a general rule, shine less brightly the further away they are;

– some offshoots are sometimes revived, which would correspond to the reaffirmation of objectives or the rehabilitation of previous actions;

– as disparate as the trajectories of these fragments may be, their center of gravity describes the trajectory that the initial intervention would have taken if it had not been fragmented, as if to preserve and extend it to its end.

8.1.3.4. *The need for an evolutionary interproject coherence*

The beauty of the wreath effect depends on the synchronization and harmony of the movements. This is why the problem of "genesis", at the margin of projects considered separately, refers to the question of the overall coherence of the elements that form the "wreath" created in this way.

It is the strategist's job to assess the capacity of the wreath for creating a true program or portfolio of interventions, on the basis of the two complementary criteria, namely:

– the degree of convergence of the respective objectives of the project/measures that compose it towards possibly encompassing objectives on a higher level.

– the capacity of the results of some to catalyze the realization of other projects, by mutually contributing to the satisfaction of their respective framework conditions, in other words, by reciprocally loosening their most significant contextual constraints.

Moreover, the job involves highlighting the synergistic or antagonistic interactions between projects, through the lens of the perception of these interactions by the actors. Indeed, as we have already pointed out, sheaf projects can, independently of their internal coherence, be perceived in contrasting ways by the different actors involved, hence the interest in detecting the alliances and conflicts they generate.

The genesis of the wreath of interventions favors a reconfiguration of the interactions and power relations between actors via an appropriate lookout or contextual surveillance function, which should be integrated into the process of global management of the action.

8.1.3.5. *Modus operandi and heuristic approach*

This wreath metaphor suggests that it is possible to isolate the first section of the trajectory of the intervention: the launching of the rocket projectile. However, there are few organizations capable of elaborating and conducting large-scale projects or measures, not considering interference with other pre-existing or future interventions. For this reason, the concept of overall management of a portfolio of measures and projects would seem to be the most appropriate, if not the only concept that could be applicable. In fact, it would be unacceptable to reach an understanding of any structuring intervention project without taking into account those that are currently being implemented. Guaranteeing a harmonious hybridization between new projects and ongoing actions is therefore a key requirement, which, if not respected, can lead to inextricable collision. Moreover, the unpredictability of the interplay of actors would exacerbate the difficulty of managing the transition between the implementation of a project and the continuation of those that preceded it.

In practice, the analysis of a problem rarely precedes the solutions that are proposed for it. This phenomenon of inversion in relation to the normative approach is linked to the fact that the idea of action is often predominant; referring to any series of problems that it is supposed to solve most often refers to the search for a justification *a posteriori*. This phenomenon contributes in part to maintaining the vagueness that characterizes most stated objectives. On the one hand, the virtual absence of a problem analysis makes it impossible to establish circumscribed objectives, since the ambiguity that remains allows these objectives to be linked *a posteriori* to certain problems. On the other hand, the degree of generality of such objectives gives them a unifying power, which it is better not to deprive oneself of, because the more circumscribed an objective is, the more likely it is to be criticized.

On the other hand, any attempt to broaden adherence to the objectives, carried out in this way, is likely to prove difficult to test. Indeed, experience shows that even massive adherence to a project is not enough to guarantee the level of involvement required for its implementation. Moreover, various studies of intervention processes have led to the counter-intuitive conclusion that there is no obvious link between the conviction of a given actor and its real involvement in the action. In other words, taking action requires the actor to be willing to devote sufficient energy to it. However, this same energy is required for other decisions or actions, which is one of the objective causes of the persistent discrepancy between discourse and reality.

This adhesion can only be crystallized through the sufficient visibility of the objective with the potential to transform convictions into active participation, giving rise to the dilemma between rapidly initiating action on ambiguous bases and reflection in an attempt to clarify its purpose and contours.

The prior search for an acceptable compromise around the intervention and its mechanisms requires active strategies. Such strategies may seek, for example, to build a coalition sufficient to converge towards this compromise that allows for "getting things done". In this case, is it not advisable to formulate a matrix of the actors involved in the reflection? From this point of view, the preliminary reflection phase should – at least when the situation becomes more complex – be based on an approach in terms of the strategic "eco-piloting" of its development. Thus, the success of a large-scale prospective study is as much a matter of reflection as of strategic action.

The possibility offered by the strategic engineering approach, to memorize and become aware of the mechanisms of reflection and decision for action, proves all the more useful when the context is contingent and when one operates in a situation of ambiguity. However, such situations never repeat, and sometimes seem to be devoid of any logic. In fact, it appears to be relatively difficult to retrace the process of negotiation-decision, to decipher its origins, or even to identify the real decision-makers, in order to understand the *terms* of this decision. However, should we take the risk of not having instruments to understand the concrete processes of reflection and action? Awareness of the intricacies of decision-making and strategic action is one of the rare possibilities available for learning and appropriating the conduct of interventions in a complex universe.

8.1.3.6. *Observance of ambiguity tolerance*

The components of the concrete context for strategic thinking and action described above would suggest some preliminary lessons.

The first of these lessons is that daily life cannot be dissociated from the future. However, the relationship between these two extremes, both often in motion, is difficult to establish: the metaphor of the "smart" missile chasing a fighter plane is an illustration of this. Contrary to our common sense in this matter, the capacity for visual piloting – both contextual and "situational" – prevails over set medium-term goals, especially if the intervention risks being called into question.

The second lesson is that the contingent and "procedural" (in terms of process) dimension of the reasoning on which the conception and conducting of action is based predominates over rational thoughts of the normative or synoptic type. In other words, the process is more important than the results, and the strategy outweighs the means. However, the strategy itself cannot be confined to a given phase, regardless of how important it may be, of a more global approach, because it is at the heart of this approach and acts as the blood which, through its branches, gives nutrients to the whole body. It is therefore the *strategic spirit* that is essential, which it is why it is necessary for there to be a massive action of awareness-raising/training with regard to this new "state of mind", the content of which remains to be defined and the pedagogy to be established.

The third lesson is that vagueness, invention, and imagination, which are inspired by the sciences of complexity, destabilize the certainties of Cartesian rationality, as well as its illusory relevance, sometimes creating uneasiness. One of the major challenges of strategic engineering consists in helping to reduce this unease by improving the intelligibility of decision-making situations and their actual dynamics.

8.1.4. *Promoting a dizzying rise towards beneficial action*

This new dimension of the strategic engineering approach is based firstly on taking into account the dialectical movement between the reconfiguration of the solution and the re-evaluation of its performance, on the basis of three macrocriteria of the strategic landscape, namely: the reference frame, the environmental scenarios, and the interplay of actors.

8.1.4.1. *The generic strategic action evaluation cycle*

The dynamic of continuous adjustments and the aggregation of partial evaluations (according to the subcriteria) into an overall evaluation or assessment of the strategic action according to each macrocriterion is progressive. For pedagogical reasons, the representation of this dynamic will initially be done in the form of feedback loops. In order to take into account the evolutionary nature of this process, we will thus use a representation that takes the form of spirals.

It is also worth mentioning that this dynamic takes shape at a finer level of observation and evaluation with regard to operational criteria. For example, with regard to strategic vigilance toward the interfaces between the action and its socio-political context, this will essentially have to focus on the salient points made up of assets and vulnerabilities. This vigilance will focus both on the severity of the critical points already identified and on the emergence of new risks or assets.

As a result, the choice of early warning indicators can be reviewed on a regular basis. This is an evolving strategic information system. It is important to pay particular attention to the changes in the evaluation reference system and the weighting system. More precisely, this must focus on monitoring the critical weighting nodes (on the criteria tree), which can potentially generate discontinuities in the overall judgement of the action, to the point of calling it into question.

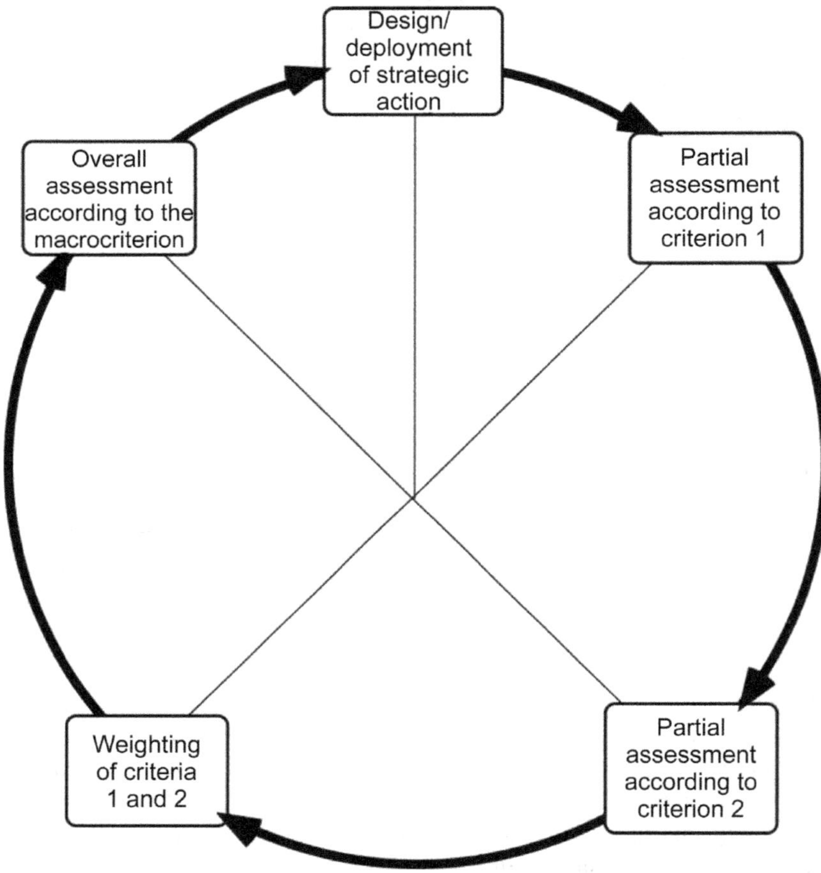

Figure 8.1. *The strategic action monitoring and evaluation cycle for each macrocriterion: a joint learning process for action and interpretation*

Any significant change in this assessment is open to interpretation. Let us suppose that a major disadvantage of the intervention proves to be unacceptable, since it had the unconditional support of an influential actor who later suffered a drastic loss in their sovereign power. Such a shift would then be accompanied by an alarming decrease in its performance score, once engaged. In this case, it would become urgent to make adjustments (reduction in the level of ambition, lobbying activity, etc.) so that this action could remain viable, even if it were to deviate from the initial course.

8.1.4.2. *The fractal progression of action monitoring-evaluation*

Drawing on the concepts of fractal geometry, the essence of which is summarized in Box 8.2, we now describe how the monitoring-evaluation dynamics of strategic action specific to each of the macrocriteria are recomposed in an emergent movement of a swirling nature. This sustainability-generating movement must be stimulated and regulated by a continuous process of *guidance,* the fulfilment of which guarantees the vitality of the enterprise and its strategic actions.

This equiangle or logarithmic spiral corresponds to the geometric shape underlying the vortex movements. It proves to be the only way to represent the processes of "continuous self-similarity".

The following quotation from Gleik (1991) illustrates our point: "A hurricane is by definition a storm of a given size. But it is people who impose this dimension on nature. In fact, people are able to observe, form phenomena ranging from the way pieces of paper swirl on a street corner to the immense cyclonic systems visible from space, that the disorder that agitates the atmosphere forms a continuum [...]. The end points of this continuum form a whole with its middle [...]. To some extent, small storms behave like big ones."

Indeed, the mind cannot easily visualize the complex result of this infinite nesting because, as observed by Mandelbrot (2010), "the structural repetitions on increasingly fine scales made possible by fractalization quite naturally open up a whole new world to those who think about shapes and explore figures from a geometric point of view".

Richardson (2007) illustrates the same phenomenon in the following terms: "Large volutes contain small volutes that draw their speed from them, and these small volutes contain smaller ones, and so on until they tend towards viscosity."

For their part, Blanchard and Waghorn (1997) argue that "to creatively manage in the midst of constant turmoil, the rule is to create strategic vortices in the workflow where it is currently stretched".

As for De Rosnay (1995), he asserts that "the organization of systems occurs at several levels, with some playing the role of building blocks in the upper levels. The system views this nesting like a set of Russian dolls. It never considers an isolated element, but always in relation to the level that precedes it [...]. Complexity can arise from simple interactions repeated myriad times from elements in constant interaction [...]. A fractal structure is constructed from repetitive patterns that reorganize themselves at different levels to produce a form that retains an analogous basic structure. The same simple principle applied systematically to each new shape creates a structure of extraordinary complexity".

Box 8.2. *Fractal geometry and vortex movement: a way to access the complexity of phenomena*

8.2. Working for sustainable viability, generating vitality

The dialectic of monitoring–evaluation–adjustment at work for each of the criteria of finality, pluralism in the face of the future, and pragmatism gives rise to a spiral progression, or rather a whirlwind-like movement. The entanglement of such modes of evolution generates a dynamic of enrichment of the system's behavior. This dynamic tends to raise the content of the evaluation by conferring on it greater subtlety, capable of transmuting it into a true meta-evaluation (of order 1). Indeed, it is less a question of evaluating the overall performance of the solution at different moments in its cycle, according to the criteria of finality, pluralism in the face of the future, and pragmatism, than of successfully integrating, over time, the processes based on these three criteria.

Thus, thanks to the swirling process, the emerging evaluation of "directivity" is superimposed and follows the more static evaluation of finalism. Likewise, the evaluation of "insight" and "agility" supplants respectively that of pluralism and pragmatism. This conceptual elevation allows us to move from a predecisional assessment of performance based on a rather "substantive" rationality, to that of a "procedural" rationality of the viability of action. It is then a question of a continuous dynamic of accomplishment of this performance.

The extension of the same process of going beyond monitoring-evaluation would make it possible to envisage a second "meta" level (of order 2), in which the notions of prosperity, durability and liveliness would be put in place as if by flocculation as emerging properties of the intervention mechanism, thus forming a new, overhanging basis for evaluation.

The ultimate synthesis of this higher-order assessment would revolve around the more essential notion of vitality. This would seem more likely to promote, in all circumstances, the accomplishment of goals and the consecration of the sustainable intervention mechanism.

8.2.1. *Gaining directivity through continuous monitoring and evaluation of finality: a dialogical spiral between voluntarism and conservatism*

The monitoring and evaluation of the finality (relevance to the vision/mission) contribute to the action being deployed. This results in a continuous series of partial evaluations (*scoring*) based on the two criteria of voluntarism and conservatism. The evaluation frame of reference evolves with the modification of the weightings attached to these two criteria, which, in turn, transforms the global evaluation of the action according to the macrocriterion of finality. The interpretation of this new evaluation of the strategic action leads to a revision of this action, and so on. The dynamic created in this way, through a succession of adjustments made "along the way" thanks to dialectically interacting criteria, takes the form of a spiral, described as "dialogical", in reference to the concept developed by Edgar Morin (2008). We will use the term "directivity" to express the dynamic of dialogical evolution in the monitoring and evaluation of finality. Continuing with the tornado metaphor, the visible manifestation of directivity would correspond to the emerging trajectory of the tornado, that is to say, to the "path" of its evolution, as observed and interpreted through the continuous accompaniment of the finality of the action.

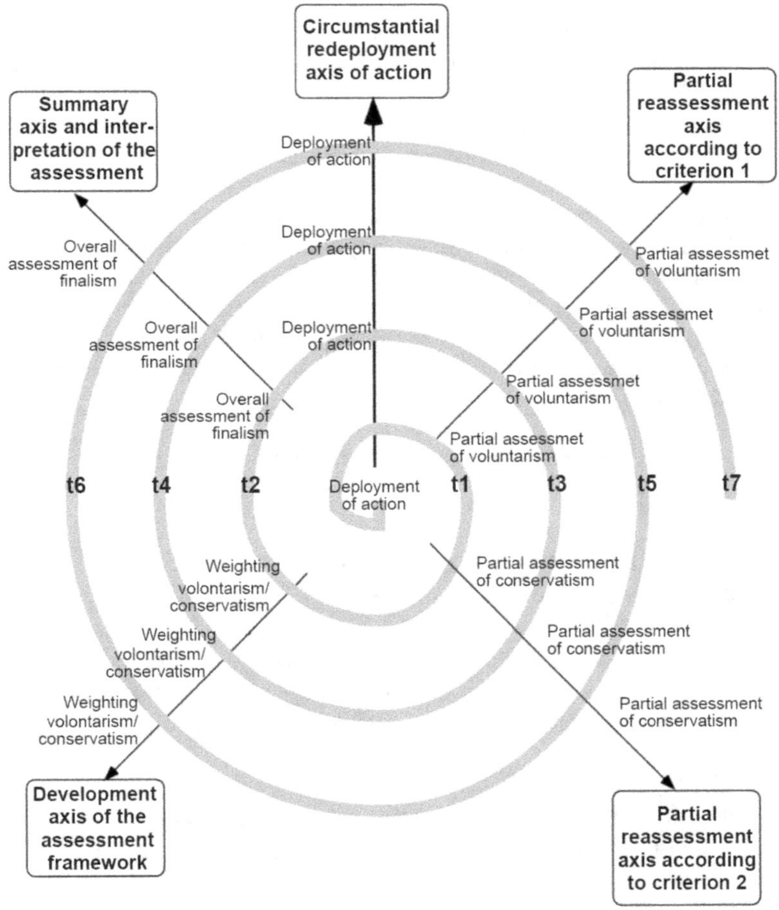

Figure 8.2. *Directivity as a dynamic of monitoring-evaluation of finalism: a dialogical spiral between voluntarism and conservatism*

The notion of directivity expresses both the ability to set the course and the capacity to exercise authority or to generate the necessary adherence to pursue it.

> The conduct of the action should tend towards a widening of the spiral around its winding axis.
>
> This widening reflects the fact that, over time, the spiral integrates a higher order of complexity, linked to the conjugated dynamics of the elementary spirals it is supposed to encompass.
>
> This metaspiral deforms as it widens. following a stretching movement along this axis which constitutes its *de facto* third dimension. It is no longer a two-dimensional spiral, but a spiral-like curve (in the form of a helix), a "vortex" in a three-dimensional space. The representation of such a curve in space seems too reductive of the dynamics thus described. Indeed, it would rather be a double-helix movement, similar to the shape of DNA.
>
> Each helix represents the trajectory of the partial evaluation of the action according to one of the criteria that dialectically compose the macrocriterion. To try to better visualize our concept, the sections of this spiral cannot be regular arcs of circles, but rather fluctuating curves. Moreover, the axis of evolution of the spiral, following the third dimension, can be brought not only to move, but also to be deformed in time. This is why the most relevant metaphor to describe this dynamic is that of a vortex or twisted tornado (composed of two entangled tornadoes).

Box 8.3. *A tornado or "vortex" motion pattern*

8.2.2. *Gaining insight and agility through the permanent regulation of pluralism and pragmatism*

As an analogy with the dynamics of directivity as the quintessence of the monitoring-evaluation of finalism, the monitoring-evaluation of pluralism in the face of the future and pragmatism also falls within the metaphor of the vortex or tornado pattern.

The dynamic of "insight" is prevalent in the case of the dialectic of idealism/ realism, which forms the pluralism oriented toward the future. It evokes the aptitude for clear-sightedness as well as wisdom.

These qualities are essential for proper monitoring with regard to the different facets of a future that is, most often than not, multifaceted and glistening.

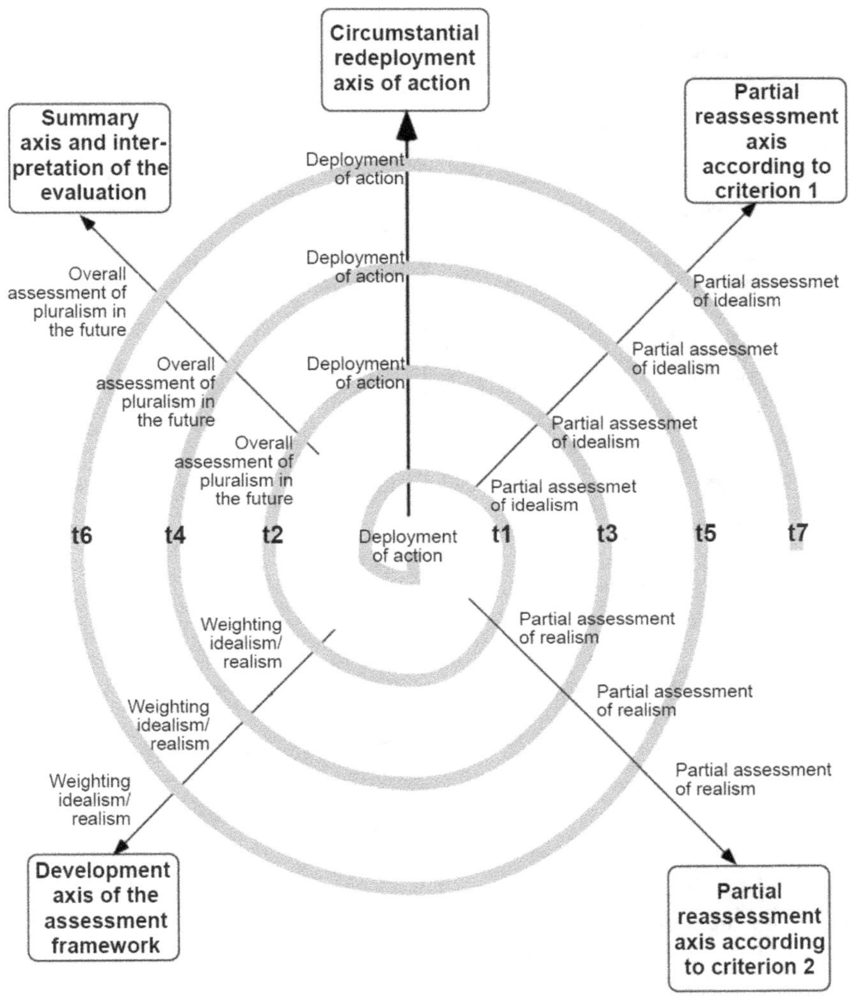

Figure 8.3. *Insight as a dynamic for monitoring-evaluating pluralism in the future: a dialogical spiral between idealism and realism*

Finally, the dynamics of "agility" emerge from the utilitarian/dialogical spiral. Opportunism arises from pragmatism. The term agility would seem richer in this case than flexibility because it reflects a less mechanistic dynamic, less intensive in investments, and above all more adapted to living systems. Indeed, it is difficult to describe foxes as "flexible", while they are more likely to be thought of as "agile", or cunning. Agility integrates two essential and synergistic properties: the flexibility of the body and the alertness of mind, which express the capacity and speed of reaction appropriate to the circumstances.

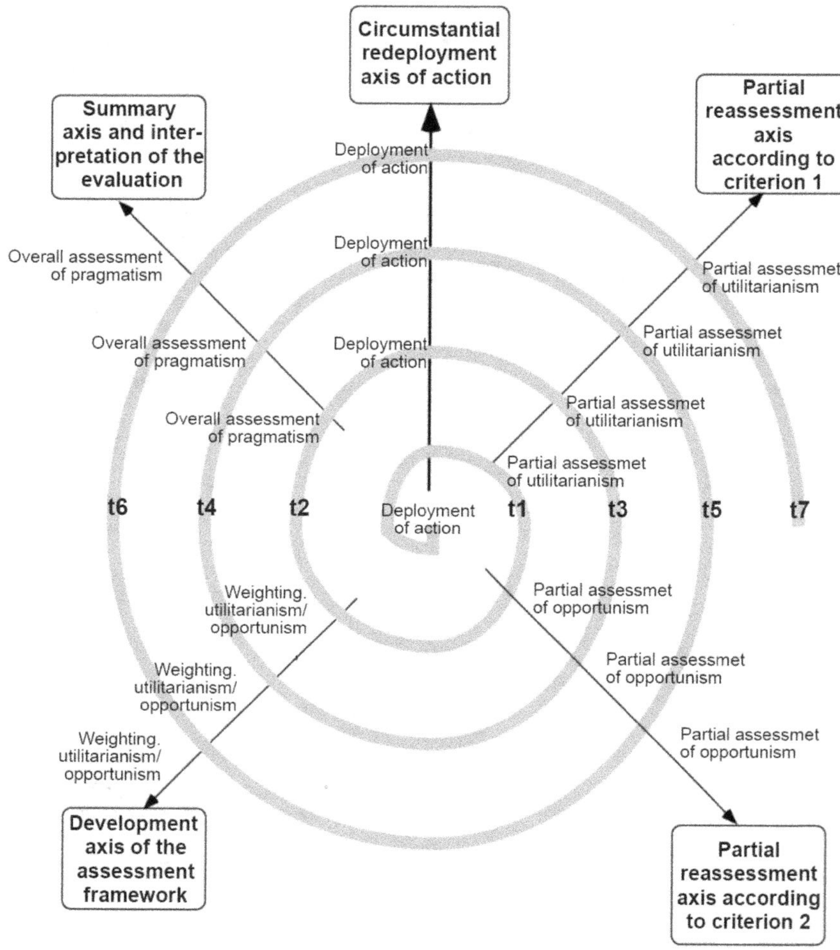

Figure 8.4. *Agility as a monitoring-evaluation dynamic of pragmatism: a dialogical spiral between utilitarianism and opportunism*

These spiral graphs (Figures 8.2 to 8.4) express how each partial evaluation can be evolutionary in terms of its results, while remaining broadly identical to itself, and thus without "distorting" itself. Moreover, the fact that these evaluations according to the three macrocriteria are expressed through the same spiral-type movement reveals their propensity to articulate and combine their progression through a phenomenon of *coalescence*.

8.2.3. *Stimulating the emergence of sustainability through a proven experience of overall performance*

The continuous expression of this complex dynamic refers to the notion of "viability of action". It involves the clearing of a path that makes it possible, as the action progresses, to monitor its overall performance, promote its achievement and benefit from its repercussions of all kinds, including those which are unexpected or *a posteriori*.

The assessment of the overall performance of the action rests on a continuous process of balancing, in the "Piagetian" sense of the term (Piaget 1971)[2], partial evaluations according to the three poles of the triad of macrocriteria: finalism, pluralism with regard to the future, and pragmatism. The dynamics associated with this three-part process of monitoring-evaluation can also be visualized through a representation comparable to a whirling motion.

Figure 8.5. *Overall performance as an evolutionary balance between finalism, pluralism, and pragmatism*

2 Piaget proposed a model of the equilibrium process resulting from the observation of perceptual regulations (Piaget 1957) and game theory. Later, he introduced a more interactionist and dialectical approach. In *The Equilibration of Cognitive Structures* (Piaget 1975), the balancing model encompasses the causes of the process and the disturbances and the means (regulations) that lead to its restructuring.

However, such a "megatornado" – based on a composition of dynamics specific to each of the poles, respectively those of directivity, insight and agility – would encompass all three tornadoes. Thus, the dynamic of monitoring-evaluation of the overall performance of the action is translated by the conjugation of movements situated at two different levels:

– one that refers to the dialogical evolution relative to each of the poles and that corresponds to the concomitant rotation of the three tornadoes on themselves;

– another that translates into a braided interlacing of previous tornadoes within the framework of a megatornado whose trajectory is itself evolving.

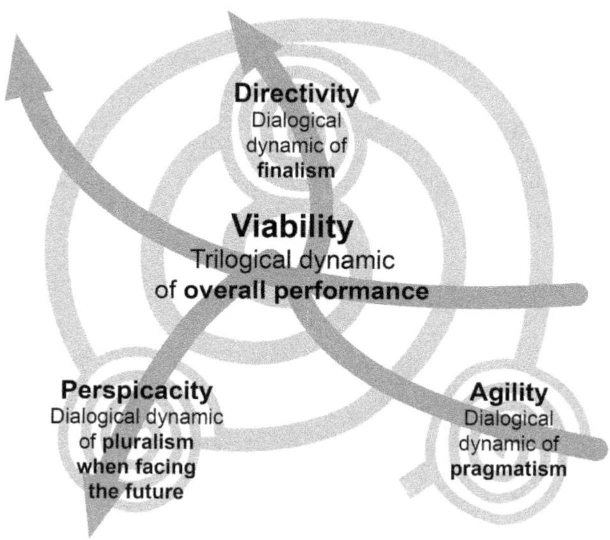

Figure 8.6. *Sustainability as a dynamic for achieving overall performance*

This new metaphor, at once broader and more complex, is constructed by an operation of nesting and fractal tensioning of movements that are similar, but not identical. In fact, the diameters of the three tornadoes reveal themselves to be different at any given moment, and evolve separately over time. In addition, the speeds of their respective whirlpools differ from one tornado pole to another and change in contrasting ways; these tornadoes can thus turn into hurricanes or even cyclones.

Thus, we are faced with a "nesting" phenomenon of the constant coupling and interweaving of dynamics repeated at different scales in the form of a braid of vortex tornadoes, which themselves are tangled. This phenomenon could in fact be applied to even finer scales, corresponding to the operational criteria of our evaluation framework.

We could also explain the recurrent search for a compromise, always unsatisfactory and temporary, between the judgments made by the different actors on the action in progress. The highlighting of this phenomenon of "unchanging scale" is characteristic of a fractal composition of the monitoring-evaluation process of the overall performance of the action.

This observation would explain why it is so difficult to apprehend and to master the dynamics of the global management of the action, in its inextricable complexity. By striving to make the dynamic of action monitoring-evaluation intelligible, the strategic engineering approach helps company or organization leaders to acquire the skills that will enable them to face uncertainty and overcome the most unexpected situations.

8.3. Ensuring the completion of the vortex elevation

As the path is cleared and the viability of the action opens up, the action becomes sustainable. The dynamics of directionality, which appeared hesitant, sometimes incoherent, ultimately turn out to be a viable path. This is a real step towards a successful outcome. At this stage, it does not matter that we have deviated from the course set *ex ante*. The essential thing is to reach a "satisfactory" situation by successive touches, by striving towards an overall performance, and by taking into account the information, the numerous feedback sources, and the interpretations accumulated and assimilated along the way.

The concept of "effectiveness", both contingent and constructed in the heat of the action, prevails over the notion of efficiency, which is rather normative and too attached to the initial objectives. In other words, and in accordance with the concepts developed by Mintzberg (1994), the result that emerges, the one that is ultimately achieved, would transcend the objectives imposed at the outset. Such a process obviously generates cognitive learning for the firm and organizational intelligence.

The mode of conducting the corresponding action would not be that of piloting, which, although referring, according to Lorino (1995), to the notions of representation and interpretation, retains a mechanistic connotation. The term "guidance" seems to be more appropriate, in reference to the metaphor of the "mountain guide" developed by Lemoigne (Avenier 1992). In fact, it expresses "the responsibility, the clearing of the path, solidarity with the rope, the discovery of the landscape, the transmission of rules of conduct according to the pitfalls, the construction of the necessary relations of trust and mutual respect, the surpassing of oneself, the existence of risks and points of no return for the rope, the extreme shared satisfaction at the end of the route, etc." (Lemoigne 1992).

8.3.1. *Gaining access to prosperity, sustainability, and vivacity as completion of directivity, perspicacity, and agility*

On the basis of continuous monitoring and evaluation of the leadership, the guidance helps to create a propensity for success and achievement, that is, to move towards the prosperity of the company. However, this prosperity is only accessible if the company has also been able to acquire two other essential properties: vivacity and sustainability.

The notion of vivacity refers to the capacity of the actor-company to understand, conceive, and react wisely, with drive and tenacity. It is this malleability to the context and circumstances of the action, according to which the firm adapts over time, that contribute to making it infallible.

The ongoing practice of insight and foresight helps to ensure the company's continued survival, in other words, its longevity and durability, despite the turbulence and pitfalls it may face.

Although sustainability, vivacity and prosperity seem to result respectively from a permanent practice of perspicacity, agility and directivity, these three first properties of the company result from the fact that they are intertwined by strengthening each other.

8.3.2. *Tending towards vitality as a sublimation of viability and a complement to prosperity, sustainability, and vivacity*

Vitality proves to be the consecration, through symbiotic fulfilment, of the prosperity of vivacity and sustainability.

"Strategists" must use the forces at their disposal, without which they would be inoperative. These forces are subtly interconnected within the company, according to an internal logic that the strategists must be able to elucidate; otherwise, they cannot reasonably guide the organization in the right direction.

Figure 8.8 summarizes the sequence of virtuous stages that the organization will need to go through to achieve a quantum leap, from the continuous achievement of its performance to the "infusion" and conscious perpetuation of an edifying life.

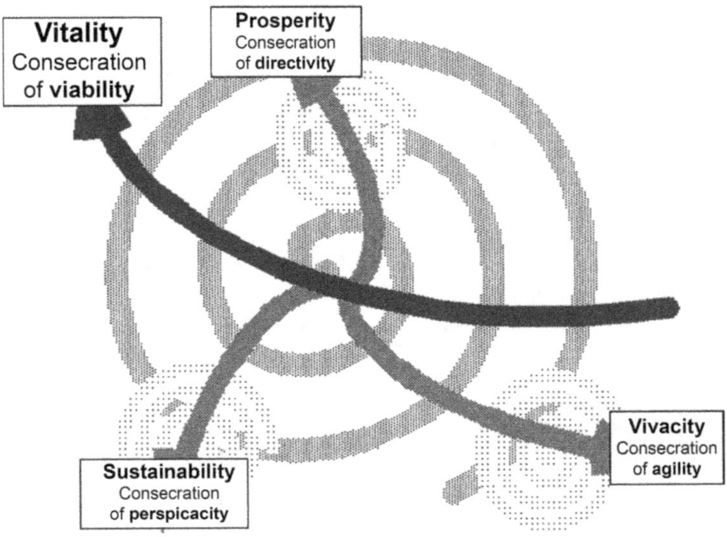

Figure 8.7. *Vitality as a consecration of viability*

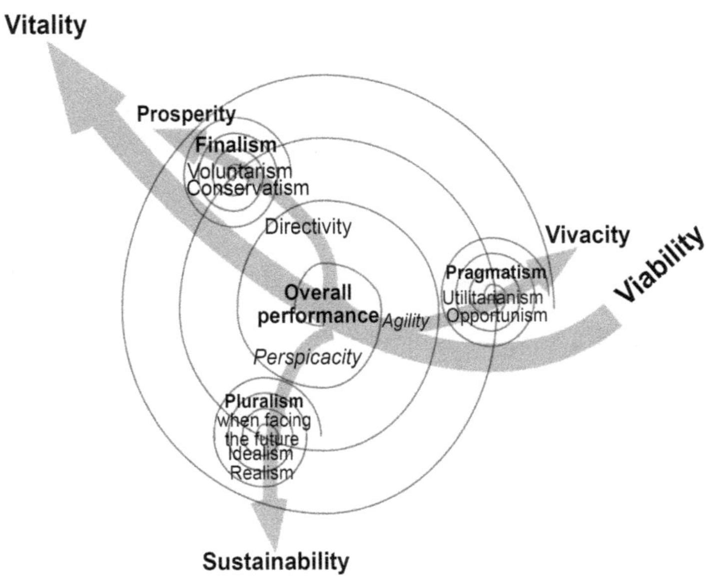

Figure 8.8. *Proven practice and inspiring experiences for sustained vitality and fulfilment*

9

Forging a Guidance System and Deploying Skills

This chapter is an essay on the possibilities of combining rigor and pragmatism in so-called "strategic" foresight. It is the result of an effort to compare the lessons learned from a wide range of methodological expertise with the concerns raised by concrete experiences in the field, both in companies and at the level of organizations or countries.

This last chapter therefore does not pretend to flesh out or extend the strategic engineering approach as such. It simply addresses the following questions: to what extent and how could we reinvigorate current practices of so-called strategic foresight in companies? In what way could such advances make it possible to initiate a process of change with the potential to facilitate the implementation of strategic engineering systems? What types of skills are required and what modes of action-training could stimulate them?

This chapter is a response to the biggest shortcomings observed in the field of strategic foresight[1], aiming identify some of the avenues for a transition of mentalities and practices towards a new "state of the art". This would be accompanied by enriched skills, in which decision-making, the search for compromises, the culture of risk taking, tolerance of ambiguity, and wise guidance play a decisive role. From this point of view, the "strategic", reified as an object – admittedly diffuse, overhanging and pervasive at the same time – could take its rightful place.

We can approach strategic foresight from different angles, depending on the situation we are confronted with, the elements of information and analysis that we

1 In reference to the work of Jean-Louis Guigou, to which we contributed (Guigou 2006).

have, and the urgency with which we must react. Each angle of attack for the approach results in a particular sequencing of all or part of the modular steps of the approach.

As an indication, three possible angles of approach to the strategic engineering approach can be noted. These correspond to different sequences of the steps involved:

– “cold” decision-making, normative in nature, of a “prospective or strategic planning” type (exploratory scenarios, proactive vision, strategic actions);

– “hot” decision-making, pragmatic in nature and of the “crisis management with pre-identified solutions” type (solutions, sets of actors, strategic reference system);

– “lukewarm” decision-making, hermeneutic and of the “search for compromise under a plurality of requirements” type (strategic references, sets of actors, value of the game).

In particular, strategic planning approaches will be examined, taking a critical and constructive look at the overall strategic engineering approach. To this end, it is necessary to:

– succeed in the delicate transition from prospective exploration to strategic action;

– guarantee an operational, long-lasting, and evolutionary articulation between foresight and strategy;

– promote organizational learning by fostering ownership of reflection and a shared understanding of the context of action and decision.

We have proposed a few contributions in order to complete the body of rigorous methods and tools we currently have. It is important to consolidate this corpus, which already articulates pragmatism and collective learning with more heuristic approaches, while maintaining synergies.

These proposals and recommendations suggest a distinctive valuation of talents and skills to facilitate access to complexity and tolerance of ambiguity. Our concern is to ensure that strategic decisions and actions benefit fully from the contributions of exploratory foresight and strategic anticipation.

These few proposals and recommendations presented in this chapter are not exhaustive and may be improved.

9.1. Reinvigorating practices of anticipation

9.1.1. *Differentiating and making normative approaches more flexible*

9.1.1.1. *Building on proven devices and processes*

The nature of the information and the way in which it is processed advocates in favor of a distinction between prospective exploration (scenario building), "visioning" (the development of a proactive vision), and strategic orientations (determination of lines of action, of a more pre-active nature).

The failure to make this distinction clearly leads to confusing scenarios with visions, or vision with strategy. The succession of these three fundamental steps in the order indicated above forms the essence of what we will qualify thereafter as a normative and/or proven approach.

9.1.1.2. *Identifying and encouraging other pragmatic inputs*

Although the procedure for sorting requirements is a privileged entry point in the normative approach, not all the information it generates is systematically used at each stage. For example, significant trends and major uncertainties are used to develop scenarios, while certain strategic aspirational and leveraging information is only used when determining the vision and major strategic directions. Thus, the sorting of available information makes it possible to ensure that certain data, while they have been generated in the initial phase of building the necessary information base, are injected into the process at the right time, without waiting for the end of the previous stage.

Other types of information, which are not part of the normative construction phase of the database, can also be reused at different stages of the strategic inspiration process. This would be the case of one or more exploratory scenarios resulting from a creative imagination that transcends the generally recommended morphological analysis method. This would also be the case of a vision proclaimed at the highest level as the *leitmotiv* of the country-actor or *driving force* of the company, or of strategic orientations obtained by a *lightning intuition*, or even a *flash of genius*. Finally, it could be the case of aspirations or requirements in terms of strategic actions, linked to strong pressure from the company's partners, the organization, the State as strategist, or public opinion.

Although the process of identifying the vision and determining strategic actions focuses on the normative process of strategic foresight, at the time that the scenarios are built, there is always an *underlying vision* that serves as a reference for the resulting commitments.

We may also be subject to pressure from actions in progress, previously referred to as "fired shots", independently of the prospective exploration exercise. Indeed, action plans have always existed, and do not wait for the formalization of scenarios to unfold and to bear their eventual fruits. Thus, whatever the results of the strategic foresight approach, this reference or these actions cannot be ignored. They are an integral part of the interventions that may or may not contribute to solidifying the orientations of the normative approach. When the normative process of strategic foresight has, for example, reached the stage of determining *ad hoc* strategic actions, these can only be envisaged in the light of the vision, or as a complement to actions already undertaken, then revised if necessary. This is why it is essential to compare the different goals of the vision with the actions envisaged in the strategic foresight process, as well as those envisaged in the context of a previous execution.

9.1.2. *Comparing and enriching the results at the different stages of the process*

9.1.2.1. *Systematization of dialectical confrontations between scenarios, vision, and actions*

It is important to consider each of the major orientations and strategic components or options and then assess their appropriateness to the different facets of the vision. A rating table is then completed to allow for a *multifaceted* assessment of each action in relation to the vision. This evaluation also makes it possible, on the condition that some simple rules for combining actions (compatibilities, synergies, etc.) and consuming resources (material, financial, human, etc.) are introduced, to assess the appropriateness not only of the actions considered in isolation, but of real action programs that are compatible with each other. The result of this confrontation offers the possibility of feedback on both the vision and the recommended actions. Indeed, it makes it possible to:

– adapt the goals of the vision in relation to their degree of activation by actions. For example, certain goals, which would not be supported by any concrete action, could be substantially adapted as a result, because they were deemed too ambitious or revised downwards in terms of their weighting-preference;

– carry out a ranking and selection of actions, redirect some of them or recompose them, by mixing them in a satisfactory manner.

Similarly, it would be desirable to confront the exploratory scenarios, as contrasting and plausible images, with the facets and purposes of the vision. Indeed, even when this vision was initially formulated in the implicit assumption of a given scenario, nothing guarantees that it will come to pass.

The approach adopted is based on a two-way movement which involves both pulling forward and deliberate orientations downwards, on the one hand, and programs and projects upwards on the other, in order to confront them. For each step, the specific results are noted. The overall approach is summarized in Figure 9.1 and is both illustrative and indicative. It is also designed to be amended along the way, in light of the progress of the previous steps and the results obtained.

1) Refinement of the deliberate directions of the vision and review of the directions emerging from existing actions:

– *Structuring of the vision and deliberate orientations*: the long-term positioning of the country or company and its major deliberate orientations (or components of the vision) are reappropriated and are schematized in the form of a tree making it possible to visualize their logic of accomplishment. Each major orientation (strategy), which corresponds to a large branch of this tree, is summarized in the form of a logical framework incorporating its own chain of objectives, as well as the main presuppositions on which its validity rests.

– *Review of existing programs and projects and highlighting of their main results, objectives, and goals*: an inventory of the programs and projects is carried out. This incorporates a list of the characteristic objectives of each component of the programs and projects. This summary is designed to facilitate the reconciliation of the objectives emerging from the projects with those arising from the deliberate strategic orientations.

2) Diagnosis of the ability of existing actions to achieve the vision:

– *Comparison of the programs and projects with the vision and its deliberate orientations*: the components of the vision affected by the programs and projects are identified. For each action, the relative importance of their impacts is assessed on a qualitative scale. This assessment is adapted according to the relative weight of the program or project concerned, as an initial approximation in proportion to its total cost and duration. A synoptic table summarizes, for each strategic orientation or major objective, the contributions or support received from the various programs or projects.

– *Examination of the adequacy of the actions to the deliberate orientations and components of the vision*: the results of the analysis are interpreted. The overall contribution of the pipeline of programs and projects to the company's or country's objectives are provided. The distribution of these contributions according to the different components (deliberate orientations) of the vision is available. Attention is drawn to those that turn out to be the least supported. In addition, a table of the synergies or contradictions of the components taken in pairs is presented. This is accompanied by a panorama reflecting the proximities and/or removals of the deliberate orientations of the vision.

3) Consolidation of orientations and refocusing of action programs:

– *Redesign and enrichment of the vision and deliberate orientations*: the critical points (unsupported components, contradictions, etc.) of the articulation between programs/projects and deliberate orientations of the vision are deepened within the framework of reflection workshops or *ad hoc* working groups which, in addition to finding solutions to the critical points identified, make it possible to broaden and deepen the appropriation of the company's strategic orientations. The tree of components and subcomponents of the vision is – if necessary – changed (moderation of certain ambitions, refinement of orientations, etc.), completed (addition of components, etc.), and restructured (modification of the hierarchy).

– *Review and improvement of the program and project pipeline*: relevant and pragmatic recommendations are formulated to improve the contribution of projects to the deliberate directions and components of the vision. The programs appear less as a catalog of projects than as a force for proposing appropriate initiatives and an instrument for channeling and harmonizing the efforts of the company or country and its partners.

Box 9.1. *An illustrative confrontational approach between the vision and strategic actions*

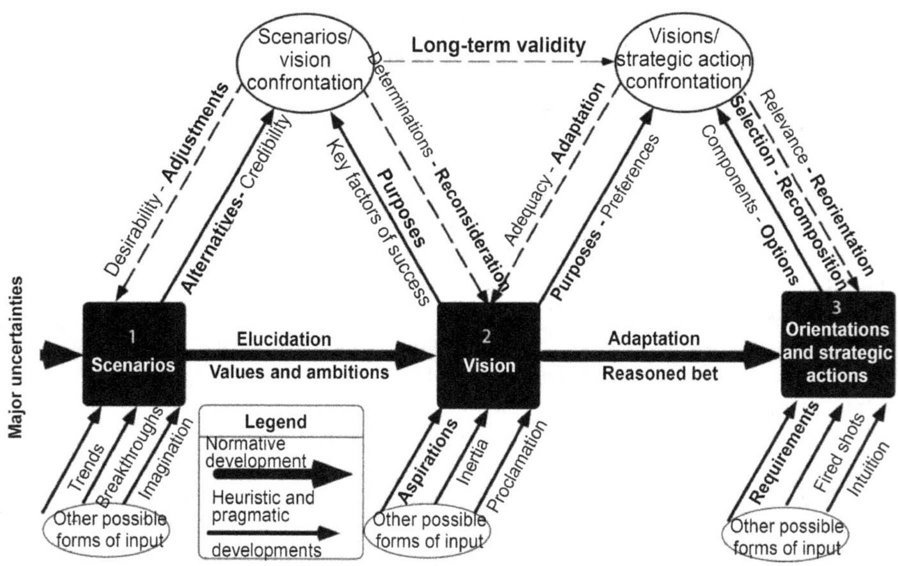

Figure 9.1. *A recursive process of strategic foresight combining a normative method and a strategic approach*

It is important to gauge the adequacy of each facet of the vision to the different scenarios (for example, through their founding assumptions). This confrontation between scenarios and vision makes it possible to both:

– make some adjustments to the scenarios, taking into account the aspirations to make them happen, the proactive nature of the vision consisting in influencing the turn or in provoking the appearance of a particular scenario;

– reconsider certain aspects of the vision or make trade-offs in favor of a particular purpose.

9.1.2.2. *Emulation of the reconciliation of results following a chain of objectives, in order to clean up the dynamics*

We note that the two types of confrontation presented form feedback loops that make it possible to progressively refine and flesh out the results obtained at each stage of the foresight or strategic planning process.

These confrontation exercises offer a significant opportunity to bounce back on actions, strategic orientations, and – to a lesser extent – on a vision or on more satisfactory scenarios based on initial solutions. Thus, these two modes of confrontation are what drive the emergence and development of the vision, orientations, and strategic actions.

However, in order to be valid and legitimate, these jurisdictions cannot be limited to the vision or actions resulting from the normative approach of strategic foresight.

This form of progressive revelation and gradual refinement also works when the initial elements of the vision and actions come from other, more pragmatic approaches, such as vision declaration, strategic intuition, or past inertia. As a result, confrontation exercises give strategic foresight a heuristic character, thanks to the use of powerful *ad hoc* tools combining simulation and optimization. These are at the heart of the strategic engineering approach. In fact, the importance of these exercises transcends the sequential order of normative determination of scenarios, vision, and strategic actions.

9.1.3. *Committing to reinvigorating the process of forward thinking or strategic planning*

There are other variants to the normative approach of strategic foresight. These variants are dictated by a concern for pragmatism. Thus, taking into account confrontation exercises diminishes the role of the normative approach and nuances the degree of rigor required for its application.

9.1.3.1. *Encouragement and support of heuristic use*

In a continuous environment, the normative approach is pre-eminent insofar as the process of successive appearances allows us to move from the exploration of the future to the determination of actions.

Where the environment is discontinuous, a multiscenario future will need to be considered. The vision, which will have different facets, can reasonably be validated only after being confronted with this pluralist and open future. Actions will have to be mixed and matched to better "fit" the many different scenarios. The appropriate mix of such a composition will only be selected after having confronted these same actions with the different facets of the vision. Finally, the more complex and uncertain the future is, the more relevant and conducive, if not essential, the confrontation exercises will be.

9.1.3.2. *Synergistic mobilization of talents and skills*

The strengthening and the art of making strategic foresight approaches operational through strategic engineering should be a major concern, both for business leaders and for those in charge of organizations or governments. This legitimate ambition should be understood rather in terms of systems, allowing the combination of methods and capacities of expertise, both methodological and specific to the themes addressed. This ambition also requires real organizational learning skills within the company, organization, or country concerned.

The initial level of expertise required for this approach concerns advice on calibration and the general architecture of the approach. At this level, the question for the organization concerned is about the selection and harmonious arrangement of the specific methods selected, according to the type of environment, the context of the reflection and the skills that can be mobilized. The purpose of this important function is in particular to establish and develop the balance between normative methods, heuristic approaches, and pragmatic contributions along the way. The realization of this function, which can be described as a "strategic engineer" function, must integrate, in particular, the emergence of the representation system underlying the strategic engineering approach. This has a triple cognitive, decision-making, and sociopolitical dimension, which enriches the global steering or general guidance function of the strategic engineering approach. However, such an approach is situated at a higher order of complexity, which could then be described as metamethodological. The qualities required in this framework are those proposed by Lorino (1995), namely: pedagogical, methodological, maieutic, and transactional.

From this point of view, the concept of the "desired end state" (DES), borrowed from military applications, proves to be fundamental. It represents the desired overall situation, when the intervention will have borne all its fruits, such as, for example, at the end of a crisis.

However, the question of the relevance, appropriateness, and robustness of the DES concept has often been raised, due to its evolutionary nature over time, and even its relative indeterminacy at certain critical moments. Thus, it is necessary to keep in perspective this notion of DES and the concept of effectiveness to which it refers. This notion is in fact all the more essential since it can be fleshed out by an approach based on "effectiveness"; that is, the satisfactory adequacy of a situation that has finally been reached.

Indeed, the political and strategic objectives are not always clear-cut and are often implicit. Moreover, when they are expressed, they are expressed in the form of recommended solutions or means to be implemented, whether as a framework for orientation or a directive, against which adequate strategic solutions can be explored and examined. In the absence of clear politico-strategic objectives or an appropriate level of comprehensiveness, a requirement-based approach should be favored. This would include not only the objectives, but also the constraints and rules in force.

In such a context of the politico-strategic decision-making process, it seems appropriate to envisage a managerial model, both conceptual and operational, which would encourage the strengthening of the distinctive competence of the strategic engineer, with particular emphasis on their singular and valuable role as "integrator of requirements".

The notion of requirement is both polysemic and concrete. Indeed, this term reflects the implicit nature of constraints, rules, and obstacles of all kinds, but also aspirations, ambitions, and political imperatives. As a result, the requirements are a challenge for the decision-maker and can be considered relatively unavoidable. The consideration of these requirements leads to a certain overexcitement by the decision-maker/strategist regarding the substance and formulation of the responses provided.

The role of requirements integrator consists of, among other things, making proposals within the framework of the politico-strategic decision-making process in terms of solutions and compromises. These proposals, most often made in a hurry, prove to be delicate and provisional. They must be prepared:

– wisely, that is, in a relevant and legitimate way in relation to political and strategic imperatives and objectives;

– quickly, which is to say, in a manner that is appropriate to the circumstances;

– in the best way possible, in other words, adapted to the constraints and rules that hold sway, or are likely to do so.

The second level of expertise involves the facilitation and conducting of specific stages of the process, including phases of gradual confrontation between scenarios, vision, and strategic actions. In addition to an excellent knowledge of the classic methods and tools of strategic foresight, this level of expertise requires a mastery of the *ad hoc* instruments of this confrontation. Through their flexibility of use and their adaptation to the logic of concrete choices, these advanced instruments offer an opportunity for constructive dialog between decision-makers and analysts. They facilitate an intimate and shared understanding of current situations or future problems. Although this second level of expertise is more technical and less complex than the previous one, it is nonetheless a privileged instrument, capable of helping to fill the many gaps, or even the "fault lines", that exist between prospective reflection and strategic action.

A third level of competence concerns so-called "content" expertise. This expertise is provided by specialists in industry, technological, legal/regulatory, and sociopolitical fields, or in the internal culture of the company or organization. It is less a question of calling on external expertise (in-depth thematic studies or hearings of specialists, etc.) than of combining a wide range of skills and expertise. Moreover, these skills can form the basis of an adequate network of contextual intelligence, strategic vigilance, and prospective awakening. Practicing exercises to compare scenarios, vision, and strategy also makes it possible to examine any unexpected – and sometimes destabilizing – situation, as measured by the reference frame coproduced and then capitalized by this network.

All of these levels and/or categories of ability and expertise outlined above must be articulated by a core of coordination and animation of the foresight and strategic anticipation process. The performance, vitality, and sustainability of this business line depend on the optimal composition of such a core group, on the methodological know-how it possesses, on its ability to generate support at the highest level, and on its capacity to cause the systems of representation of the participants to evolve at the different stages of the reflection process. Everything is subject to a real business project, combining capacity development and the production of results, using a network approach.

9.1.3.3. *The sequence of the approach in successive nested cycles*

One way of linking these different skills by combining a normative (sequential) and pragmatic (recursive) approach is to envisage a path in the form of nested cycles, the principles of which we propose to outline (Figure 9.2).

An initial cycle of reflection of a pilot nature can be carried out in an accelerated manner. It involves examining the entire process, taking into account the work context, in order to test the skills and information available, and assessing the possibilities for mobilization at the higher level.

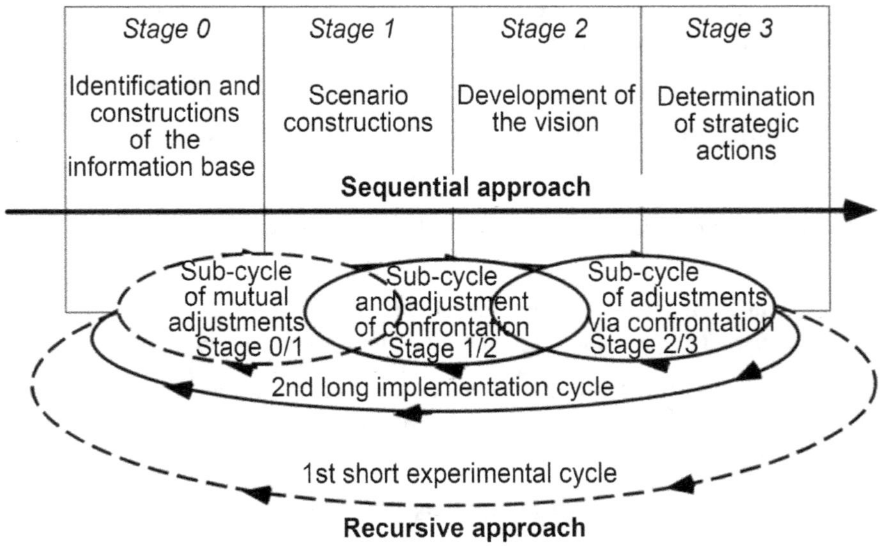

Figure 9.2. *Sequential and recursive approach*

This cycle should give sufficient visibility to the reflection process and put the methodological instruments into perspective within the framework of a global approach. It is also intended to produce the first elements of content, in terms of the preliminary results of the approach. By shedding light on the overall formatting, and even on the appropriateness of the complete approach of strategic foresight, this initial cycle offers the coordination and facilitation core a valuable opportunity to collaborate – and to benefit from a competitive edge – both at the methodological level and in terms of the content of the results that may potentially emerge from it. The common practice of seminars launching a strategic outlook approach, including *ad hoc* workshops (factors of change, preconceived ideas, skills trees, etc.) is done to accomplish this objective. However, in most cases, this would not be sufficient to ensure satisfactory visibility or to constitute a decisive step towards a successful reflection-action process.

The second cycle involves the implementation of the in-depth approach of prospective or strategic anticipation. Starting from the identification and construction of the basis, it is grounded in the sequence of steps relating to the construction of scenarios, the creation of the vision, and the determination of strategic orientations and actions. This second cycle can incorporate subcycles for comparing and adjusting the results of consecutive steps between the scenarios and the vision, and between the vision and the strategic orientations/actions. Such adjustments may lead to a gradual revision of the overall approach. The easier it is to understand such amendments, the easier it will be to understand that the pilot cycle will already have been completed.

9.2. Making the process more fluid and sustainable through deployment plans and projects

There are two types of institutions directly involved in strategic foresight approaches:

– "visioning" and strategic steering bodies within the organization. These are generally constituted in the form of committees, higher councils, etc.;

– the foresight or strategic planning body, which may, depending on the case, take the form of a cell and/or a think tank, a directorate, or a department.

9.2.1. *Considering the main structures involved*

Despite a certain decrease in both the number of staff and the powers of the strategic planning body, these two types of bodies have relatively substantial resources, giving them a certain weight.

They are engaged, respectively, in the management of continuous processes:

– of politico-strategic orientation and decision-making;

– of strategic planning and action programming.

The creation of a prospective reflection system is carried out quite lightly, in order to animate the strategic anticipation process within the organization. It is set up in the form of a multidisciplinary task force or through a cell, sometimes temporary. However, the degree of institutionalization of such a system is generally not comparable to that of steering or planning bodies.

9.2.2. *Strengthening institutional anchoring and organizational learning*

The three essential phases of the anticipation and strategic planning process are, in essence:

– the coordination of preliminary context analysis studies and the construction of exploratory scenarios;

– the choice of the vision and the establishing of the main development axes;

– the development of the strategic direction framework and multi-year rolling action programs.

The apparent continuity of this approach is most often translated by an effective discontinuity due to the changing of a main actor. This discontinuity can be compared to that of a river that has been diverted from its course, imposing an artificial meandering course. This metaphor partially explains the relative lack of fluidity that can be observed when moving from prospective reflection centered on the construction of scenarios, to the determination of the vision, and then of strategic actions. The question may therefore be asked whether this virtual dam could not be attributed to the very content of these last two phases of the strategic foresight process. To what extent could one suggest a different conception of the strategic foresight process that would make this passage more fluid, even if it meant modifying the content of the last two phases, or even reconsidering – if necessary – the object of this approach?

9.2.3. *Preserving the logic and prerogatives of the preceding parts*

The vision determination phase must be in line with the fundamental orientation and strategic steering functions devolved to the bodies concerned (executive committees, boards of directors, etc.). These functions, which are listed in the successive boxes in the first line of Figure 9.3, include, upstream of the vision, monitoring of the prospective reflection work. In particular, they lead to the validation of exploratory scenarios. Once adopted, this vision must be sufficiently declared and disseminated to be appropriate. It will only be meaningful if it is adapted to changes in the external environment and to the circumstances of the strategic action, but also debated and reaffirmed as a consequence. In the same way, it is necessary to put into perspective the functions of the planning organization in relation to the strategic foresight approach (the third line of the figure).

These functions include the development of the strategic direction framework and "rolling multi-year programs" of action. However, this phase, which is considered to be part of the strategic foresight process, must be based on a satisfactory appropriation and correct application of the vision created by the steering committee, while taking into account ongoing programs. In order to remain relevant, these strategic orientations and actions will be periodically redirected or adjusted according to how they are actually implemented and any changes made to the vision.

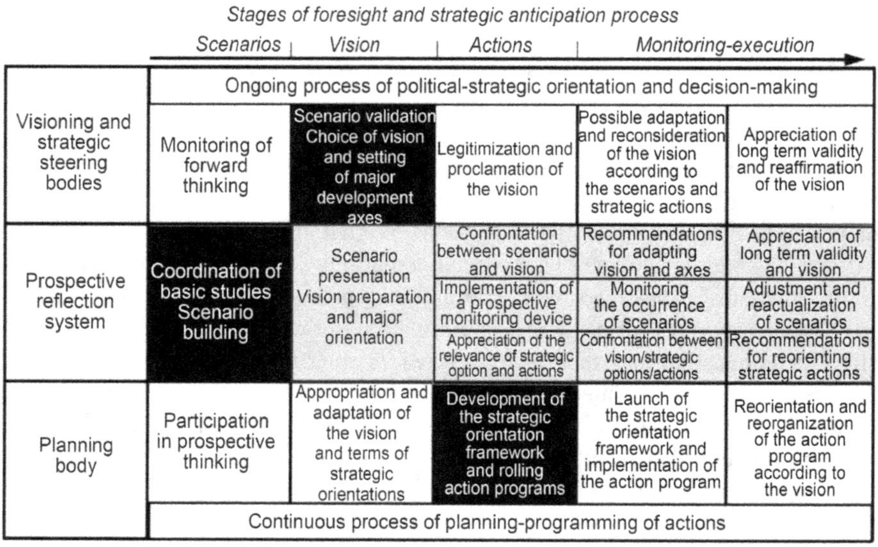

Figure 9.3. *Connection between the institutional mechanism and the foresight approach – strategic anticipation, according to the bodies involved*

9.2.4. *Opting for a strengthened approach capable of percolating through the organization*

With regard to the foresight reflection mechanism (the blue part of the second line of Figure 9.3), which is at the heart of the work of exploring the future as embodied in the scenarios, it would be advisable to give it a more modest role during the visioning and strategic action development phases, to avoid creating mistrust or even opposition in the competent authorities. On the other hand, it would seem feasible for this mechanism to focus on a core set of missions and competencies, which should be revisited. It incorporates two main functions in particular:

– circumstantial support for the functions of the strategic steering and planning institutions, while respecting their own prerogatives. This type of support would essentially consist of exercises of confrontation and reconciliation between scenarios and vision and between vision and strategic actions;

– prospective surveillance and monitoring of the advent of scenarios.

These two major functions can be considered fundamental, in that they are in the natural wake of the logic (river bed) of reflection that prevailed for the construction of the scenarios. They also contribute to the success of the visioning and planning functions. Finally, they remain underdeveloped and can only be implemented through a specific mechanism. These functions imply both a salutary restriction and an unprecedented extension of the usual foresight and strategic anticipation approach. Implementing them, as part of a renewed approach, would offer two opportunities. The first would be to enhance and make better use of the distinctive capacities and skills of the thinking system, especially when they prove to be irreplaceable. The second would be to appropriate and perpetuate the system, to update the scenarios, adapt the vision, and reorient strategic actions.

These considerations, which are essentially technical and institutional in nature, lead us to consider mastering the methods and tools for pragmatic confrontation-conciliation between scenarios and vision, or between vision and strategic actions, as essential. Although counter-intuitive, this precept nonetheless encourages us to minimize or reconsider the conception of current approaches to strategic foresight. Indeed, the latter remain based on a normative and linear process of sequential elaboration of the different objects of this confrontation (the vision, orientations, and strategic actions) considered in isolation. However, these different objects interfere, or even collide, at critical moments.

9.2.5. *The case of a project to strengthen technical-institutional capacities in strategic foresight*

In order to illustrate our point, we will present, partially and in summary, an example of a development project in the field of strategic foresight. It was a project implemented at country level, with technical assistance and funding from the United Nations (UN). Box 9.2 is limited to the main components of the first column of its "logical framework" (see section 3.2.3).

Development objective: to reinforce the capacities of anticipation and strategic management of development through the creation of a prospective and concerted frame of reference in order to define a consensual long-term vision of the Organization.

Objective 1: gather and elaborate on all the information necessary for reflection.

– *Expected results*:

- factors and mechanisms of evolution,
- primary trends and seeds of change,
- inventory, concerns and intentions,
- bibliography and reconstruction of a documentary fund of basic information.

– *Activity: preliminary studies:*

- retrospective studies,
- aspiration surveys,
- synthesis and presentation of the results of the survey and retrospective studies,
- inventory of other available studies.

Objective 2: understand the main determinants of the evolution of the Organization as a system.

– *Expected results:*

- visual representation of the "organization system",
- understanding the dynamics of the system and the role of the factors,
- identification of major themes and key questions of foresight,
- reference terms for complementary studies (focus groups, thematic analyses).

– *Activity: structural analysis of the "organization system"*:

- review of the factors and variables of evolution,
- reconstruction of system interactions,
- processing and interpretation of results.

Objective 3: explore possible contexts for the development of the Organization within the 2025 horizon.

– *Expected results*:

- proposal of relevant, coherent, and plausible scenarios for the global context of the Organization's development over the prospective horizon, each scenario consisting of a cascade of hypotheses about the future,
- key messages of foresight and critical requirements for the strategy,
- specifications for complementary strategic studies.

– *Activity: development of scenarios*:

- construction of scenarios by theme as combinations of hypotheses on key issues,
- development of global scenarios by aggregating thematic scenarios,
- synthesis and presentation of exploratory scenarios (thematic and global).

Objective 4: develop a unifying and mobilizing vision to shape the future of the Organization.

– *Expected results*:

- a fundamental development option and the main ambitions of the Organization on the prospective horizon,

- the major challenges to be taken up and the main assets and risks attached to the vision,

- the reference terms for the deepening of ambitions.

– *Activity: determination of the strategic vision:*

- identification and deepening of the decisive stakes for the Organization,

- determination of the vision as a strategic surpassing and voluntarist positioning of the Organization in its future context,

- analysis of the relevance and robustness of the vision in relation to the scenarios.

Objective 5: provide the Organization with a "reference framework" that will help to clarify the structures and guide strategic action in consultation with stakeholders (strategic reference framework).

– *Expected results*:

- long-term strategic referential of the Organization, connecting the exploratory scenarios and the stakes, the vision and ambitions, as well as the challenges and the axes of breakthroughs,

- a first sketch of an integrated development framework, including the positioning of the main actors,

- the reference terms for the deepening of the axes of breakthroughs and/or the positioning of certain actors.

– *Activity: specification of the foundations of the strategic action*:

- declaring the major strategic orientations and determining the guiding principles to fulfill the vision and ambitions for 2025,

- analysis of the actors' games around the vision,

- identification of strategic breakthroughs to address obstacles and challenges.

Objective 6: propose a system for the sustainability of the foresight and the "operationalization" of the strategic referential.

– *Expected results*:

- proposal of a methodological and institutional framework to perpetuate the function of strategic foresight,

- a grid for assessing the relevance of strategic actions in relation to the vision,

- recommendations for taking into account the vision and long-term ambitions in strategic actions,

- proposed framework of strategic direction documents.

– *Activity: monitoring/evaluation of strategic foresight*:

- design and formalization of a permanent function of strategic anticipation and vigilance,

- periodic confrontation between vision and strategic actions (policies, strategies, programs, major projects, etc.),

- contribution to the preparation of major arbitrations in reference to the vision,

- proposal of guidelines for developing and/or influencing strategic actions.

Box 9.2. *Project objectives, activities, and expected results*

9.3. Practicing, integrating, and appropriating the strategic engineering approach

From a conceptual point of view, the strategic engineering approach makes it possible to move from a normative-type model of efficiency to a relatively new form of pragmatism. This form emphasizes the viability and effectiveness of actions, partnerships, and their accomplishment. Such pragmatism is accompanied by a new type of innovation of an integrative and societal nature. This is based on a permanent tension between: (1) the different levels of analysis, (2) the focus of initiatives at the same level, and (3) each fragment of action and its immediate context.

9.3.1. *Thinking of organization and action as holomorphic*

A holomorphic field is by definition made up of entities with similar characteristics and properties at different scales. This likeness is not perfect, due to the fact that the parts in question are autonomous and differentiated. They reflect the essence of an undivided whole, which depends both on the underlying model of strategic engineering and on the importance of the contexts, according to the scale. This is why the apparent dissonances that could result from them paradoxically conceal a double coherence attributable to the mode of representation and to the content.

9.3.1.1. *A fractal geometry from the whole to the unit and/or from the unit to the whole*

Representing the strategic landscape through nested fractal geometry frameworks is an adequate way to assimilate the holomorphic character of the organization.

The consideration of the conditions and manifestations of actions in their context diffracts, at different levels, so that the global model of strategic engineering applies to different scales of observation and to different levels of granularity or of fragmentation of analysis.

Thus, we can introduce the notion of an "elementary entity of representation" which would result from the association of a basic component of the action to its inseparable immediate context or "contextual neighborhood". This basic unit, which we describe as a "praxeon", represents the elementary cell of the model for decision and enlightened guidance of strategic action. Thus, the holomorphic nature of the action in its environment would tend to favor a development by contagion, abundance, and percolation of the process of penetration and dissemination of strategic engineering at different scales. This process takes place through "strategic instillations" that are contingent on their specific contexts and on their own temporalities. We are seeing an abundance of "glocal" centers of initiative, articulating both global and local polarities. This articulation is made concrete through a plurality of specific subsidiarities that take place at relevant levels, appropriate to the way in which the situations or problems concerned are dealt with. Although these centers may be autonomous, they cannot be considered as independent, because their paths of progression are intertwined, shaped and mutually reinforcing, in an evolutionary manner.

9.3.1.2. *Situational management for informed guidance*

Due to its holomorphic nature, the strategic engineering approach is rather catalytic, in that it contributes to creating favorable conditions for the development of the whole, in this case of the strategic landscape as a whole, by favoring the freedom of each actor. Moreover, strategic action can benefit from being stimulated by a set of reasoned incentives. These are targeted at crucial points or challenges, which mutually echo each other, as if to reinforce the action as a whole. It is a "situational" type of management. It is carried out through incentives and consultation at different levels, with a permanent "looping" around the consequences of the components of the actions being tested.

In this way, politico-strategic action acquires legitimacy and credibility based on the exemplary nature and experience that the process of enlightened guidance confers on it.

9.3.2. *Organizing conferences, specific training courses, or on-site strategic engineering projects in a modular manner*

These training courses are vectors for the dissemination and deepening of anticipation and strategic engineering issues, and meet three pedagogical objectives. These are accompanied by a variety of modules covering the different aspects of the approach.

9.3.2.1. *A proposal of pedagogical objectives and heuristics*

– To constitute an opportunity for improvement and deepening in the preparation of the future and innovative anticipation, as well as in the problems of decision and politico-strategic arbitration.

– To specify the principles and concepts of foresight and choice in an uncertain future and/or uncertain environment, the philosophies that underlie them, as well as the methods and tools they induce, emphasizing their virtues and requirements on the operational level, as well as the pitfalls inherent in methodological clichés or incongruous practices.

– To promote the appropriation of the principles of implementation and facilitation of devices, as well as approaches in terms of innovation and/or change, through the presentation of a structured approach, the discussion of proven practices illustrated with cases, and the realization by participants of tutorials in subgroups.

9.3.2.2. *Modules constituting the strategic engineering approach*

This training includes in substance the main modules of the strategic engineering approach, including: the construction of prospective scenarios, the anticipation of the situation and its implications, the mapping of the actors' interplay and their power relations, the determination of a final state based on the vision, the creation and evaluation of strategic options, the systemic anticipation of risks, the choice and recommendation of an intervention project in the form of a logical framework, the reference frame for monitoring and evaluation, and the circumstantial regulation of the action.

9.3.2.3. À la carte *programming of targeted conference cycles*

As an indication, and not exhaustively, the various aspects or fragments covering the strategic engineering approach and likely to be addressed in the context of conferences or training are presented below. These are judiciously grouped and synthesized, in consultation with the partners/beneficiaries, to give rise to a wide variety of modules. These different modules can constitute a range of themes that can be combined for the organization of *ad hoc* conference-debates, an indicative list of which is proposed here:

1) "Construction and staging of prospective scenarios": exploring future contexts in an effort to take advantage of them.

2) "Discernment, impregnation, and regulation of the sociopolitical landscape": deciphering the interplay of actors to instill a dynamic.

3) "Visualization of the angles of attack of 'bastions' to be conquered": revealing and resolving the issues in order to focus on the major challenges.

4) "Determination of the main orientations and breakdown of the objectives": elucidating one's aspirations and ambitions to give meaning to the action.

5) "Design and preselection of mutually exclusive actions": outlining a range of options to assess their relevance.

6) "Adoption of a strategic position adapted to the climate of uncertainty": assessing the decision-making logic to adopt a particular approach.

7) "Judicious composition and choice of a satisfactory mixed solution": recombining the options to decide on the best strategic mix.

8) "Anticipation of risks and judicious recommendation of precautionary measures": apprehending possible bankruptcies in order to prepare for any eventuality.

9) "Formulation of the chosen solution and its conditions and success factors": synthesizing the development project in a coherent way.

10) "Continuous evaluation and circumstantial regulation of the action": implementing a process of guidance and change management.

11) "Fruitful redeployment and access to organizational fulfilment": instilling collective vitality in order to strive for full development.

12) "Feedback, testimonials, debates and perspective": organizing and deploying a strategic engineering system.

The order of the first 12 modules given here represents the natural progression of the overall strategic engineering approach. However, this order may be modified, provided that a satisfactory level of appropriation is guaranteed.

These first modules can, if necessary, be completed "*à la carte*" by about ten additional modules for further study and/or improvement in strategic engineering, of which an indicative list is as follows:

1) "Wise exploration and evaluation of proactive deposits": identifying glass ceilings to open up new horizons.

2) "Injection of disruptions or surprises and creative generation of the unknown or the unusual": "pushing the envelope" of scenarios or options that are rather conformist.

3) "Extension, aggregation, and subsuming of available results and/or reflections in progress": amplifying and diversifying the space of the future and/or initiatives.

4) "Appreciation of the implicit power of sublimating audacity by exploit": identifying and conjugating the cardinal strategies of uncertainty.

5) "Circumspect detection and regulation of strategic risks": curbing crucial risks in order to avoid fatal errors.

6) "Insertion at different scales of rather discriminating descriptive components": fleshing out the scenarios in order to highlight their inspiring and stimulating character.

7) "Multilevel configuration of the chain of incidences and conditions for carrying out the action": drawing up requirements and preconditions necessary for the finalization of the action.

8) "Stratification of a posture of meta-order by stacking intersecting layers of uncertainty": hybridizing strategic postures respectively adapted to situations of uncertainty.

9) "Review of and perspective on all the preceding points": scanning the different possible areas for further study.

10) "Custom-made proposal of courses based on an adequate arrangement of particular points": designing a substantial program of in-depth strategic engineering.

The choice and the order of the ten complementary modules are decided upon with the goal of effective learning and significant appropriation of their contributions by the companies or institutions concerned. Indeed, these modules constitute as many methodological and practical wastelands to be developed and exploited within the framework of innovative strategic reflections. Some of these avenues for deepening strategic engineering can, if necessary, be the subject of *ad hoc* research-action *in situ*.

9.3.2.4. *Higher education in "exploratory foresight and strategic anticipation"*

Courses in this area can be taught at universities or in major schools. They could take place over approximately 30 hours, spread over five sessions of six hours. Each session would be devoted to one of the five modules of the indicative program presented in Box 9.3. The course would include, in particular, the conducting and subsequent monitoring of case studies in the form of tutorials in subgroups. These would be constituted by co-opting students from a list of topics proposed by the whole class and then selected according to their interest for the class as a whole. For this purpose, an average of two to three hours maximum per day module of training would be devoted to these workshops in subgroups.

These tutorials would be followed by a monitoring and evaluation session, in turn followed by an oral report from each subgroup in the presence of their peers. These results would be summarized in reports, duly evaluated as part of a continuous monitoring process.

Exploring future contexts and posing fundamental issues

– Introduction to exploratory foresight.

– Sorting of variables and identification of major uncertainties.

– Sequence of the scenario-building method.

– Transcription of scenarios into hypothesis trees about the future.

Adopting a proactive vision and take advantage of actors' games

– Identification and appreciation of critical issues and challenges.

– Decisive arbitrations and building a vision in terms of orientations.

– Understanding of the actors' interplay and anticipation of their effects.

Developing a range of strategic structuring options

– Identification of areas of intervention and strategic levers.

– Recommendation of methods of action for each lever.

– Determination of strategic option concepts.

Assessing options against scenarios and directions

– Consolidation of the evaluation framework in the form of a hierarchy of criteria.

– Clarification of preferences and weighting of criteria.

– Multicriteria comparison of options and sensitivity analysis.

Organizing and succeeding in a process of prospective and strategic thinking

– In-depth study of the recommendation of mixed solutions.

– Organization, coordination, animation, and monitoring of the process.

– Deployment of a system equipped with anticipation and feedback tools.

– Conclusion and perspective towards strategic engineering.

Box 9.3. *Indicative program of the "exploratory foresight and strategic anticipation" course*

It is important to emphasize the relatively novel nature of the approach proposed, for most students, as well as the diversity and breadth of the fields covered by this work. The work is satisfactory, both from the point of view of assimilation of the proposed approach and the content of the results obtained; and this in relatively difficult conditions, due to the complexity of the subject matter, the demand for content, the time allowed for the submission of reports and the quota of hours devoted to the workshops.

While this work requires substantial effort on the part of the participants, it allows each person to take ownership of the proposed approach in his or her own way and then to experiment with it in the exercise of his or her professional activities.

9.3.2.5. *Introductory seminars, advanced workshops, and operational workcamps*

All or part of the modules presented above can give rise to introductory or training sessions, in-depth seminars and workshops, and work sites in an operational environment, but also to higher education at the level of a master's degree.

The initiation and training cycles or sessions aim to present the integrated approach and the software approach, in order to aid in deciding wisely, intervening in the best possible way, and guiding politico-strategic action. This consists in reviewing all or part of the modules presented above, at a rate of two to four per day. As an example, the indicative program for a session of two consecutive days would include a synthetic and accelerated presentation of the academic course program proposed above.

The in-depth and advanced seminars and workshops would involve intensive training and the practice of strategic anticipation, to facilitate collective learning and appreciation of its scope.

These seminar-workshops would include, in particular, the development of the approach and the *in situ* experimentation of the methods used through the study of a concrete case in groups with a variable arrangement, selected according to its politico-strategic relevance and its pedagogical interest. This case study would give rise to the creation of a synthesis of the work accomplished and its most important results. It would also include a presentation of the conclusions of the study and a discussion, in the form of round-table discussions, of the primary lessons of this learning process.

Strategic engineering projects in an operational environment involve the in-depth and operational implementation of strategic anticipation engineering within a company or organization. These should be carried out according to specifications to be determined in the situation. Each project should be carried out through a custom-made program with a schedule that may include several sessions lasting about a week. This may possibly be made to include technical assistance or advice, and may be subject to the provision of specialized software.

9.3.3. *Promoting and developing advanced strategic engineering instruments*

The multiple fields and areas of application of foresight, the growing concern for strategic action, as well as the demand for support for reflection mechanisms, have led us to develop, within the framework of intellectual partnerships, three new concepts accompanied by new logical instruments, in the form of demonstrators, in the field of strategic engineering.

9.3.3.1. *Specialized tools to support the approach*

The purpose of these computerized methods is to respond to the multiple requirements that have arisen in the areas of advice, services, and strategic interventions, such as prospective exploration and scenario development, preparation of major trade-offs and deployment of strategies, intelligence of the contexts of action and sociopolitical vigilance.

These three instruments[2], previously presented as boxes, have been designed respectively to:

– facilitate the playful construction and structured presentation of exploratory scenarios and strategic options (Morpholudis, see Box 2.2);

– strengthen the capacity for strategic intelligence and trade-off intelligence (Astris, see Box 7.1);

– provide assistance on how best to decide on composite solutions in a complex and uncertain universe (Simoptis, see Box 6.1).

They also make it possible to complete, enrich, and surpass the multitude of tools available in terms of "strategic" foresight. Finally, they contribute to cementing and strengthening the crucial link between foresight, strategy, and monitoring.

9.3.3.2. *A heuristic system integrating these tools*

These three instruments are also integrated within the framework of a heuristic methodological system: PraxéoLab. This is accompanied by an integrated, experimental software program for strategic engineering, PraxéoGraphis. This simulation and strategic decision support demonstrator was initiated on behalf of the

2 Developed by Jacques Arcade (previously Proaxis).

French defense ministry, based on our original concept, cocontracted with Atos Origin. As shown in Box 8.1, PraxéoGraphis is designed as a host structure integrating the three specific instruments, Morpholudis, Simoptis, and Astris, mentioned above.

9.3.3.3. Ad hoc *teams of experienced stakeholders*

The conference, teaching, and action-training cycles are coordinated and led by small *ad hoc* teams of strategic engineering specialists. These teams can be constituted according to the concerns of the client-participants and the nature of the interventions. These training cycles generally meet the three pedagogical objectives set out in section 9.3.2. They are accompanied by a variety of modules covering the various aspects of the approach.

Conclusion

Strategic engineering provides learning institutions with a methodological framework that allows them to conceptualize large-scale actions and assess their impact. It also allows them to simulate the immersion of these interventions in their strategic landscape, in order to improve them and attempt to regulate them.

A twofold relationship thus emerges and manifests itself, according to which the organism impacts its situational context and leads it to transform itself, encouraging it to "accommodate" and evolve. Moreover, the hold exerted by this context pushes the organization to "assimilate" itself to such context, by modifying its own patterns or by diversifying its repertoire of actions. This process, similar to that of Piaget, would lead to a progressive and relatively asymmetrical adaptation of the organization and its context. After numerous adaptations, this would lead to a partial and temporary "balancing" of this twofold movement of accommodation and assimilation, which manifests itself at the interface between the organization and its context. This constitutes the ambient environment or frame of reference for designing and deploying strategic actions. This "environment" includes, among other things, the intuitive space of possible initiatives, characteristic of their field of operation. It proves to be consubstantial with the organization and its collective of actors-strategists. These will be able to project themselves in the interest of the organization by immersing themselves in this environment or "world". They will therefore be able to act, in full awareness and knowledge of their creative abilities or the scope of their common actions.

This world "for itself" (in consciousness) created in this way does not exist outside the collective. It cannot be considered as simple decor, a provider of exogenous representations. It is essentially the product of the "situated perspective" that is offered to the strategists involved in this cognitive adventure, and provokes a quest for meaning and significance on their part. It is therefore this sharing of "added meaning" and the construction of common sense that ultimately confers on this collective an identity of solidarity that is destined to be consolidated and extended within the enterprise or organization.

By virtue of the principles of projective geometry and the laws of perspective, and in accordance with Leibnitz's theory of monadology, our reflection refers to the duality between the "one" of the monad, and its appearance in the form of a constellation (the collective) formed from its different constituent "ones". More fundamentally, according to Varela (2017), this milieu would benefit from being "enacted" by our collective of actors-actants, in order to generate wisdom within it, even inexpressible wisdom. Under these conditions, organization would become capable of testing and animating this world. through experience lived in the present. According to Merleau-Ponty (1976) and Husserl (1985), this "experimentation" would be essentially phenomenological in nature. Beyond the ecosystem intelligence necessary for strategic engineering, the world created and animated by the community of actors in this way confers on the organization a praxeological type of wisdom, which tends to rise to the "demiurgic" level.

However, the movement spurred on by such overachievement requires more attention to ethics on the part of companies, organizations, and decision-making bodies.

Therefore, with the idea of carrying out judicious extensions of the strategic engineering approach in mind, we can participate in the emergence of new paradigms. These paradigms could prove to be the bearers of inclusion, resilience, preservation of nature, and respect for humanity and life in general. The approach to achieve this requires, at the very least, the designing and development of an appropriate reading and comprehension grid. This will have to prove inventive, resilient, and more or less qualitative. Creating this grid will require the participation of various collectives and communities. The goal is to comprehend and take on critical challenges, with resonance at the global level, at different scales.

Such a design would force us to take better advantage of the irreducible complexity of today's world, in an attempt to break through its inertia, which is predominantly technical and "economic". We must ensure that our world of tomorrow opens up to new horizons that are comforting and more engaging for our communities in action and for the new humanity that is to come!

Postface

When, at the end of the 1990s, the French Ministry of Defense began studies on strategic decision support, as the policy officer of the Délégation aux affaires stratégiques (Delegation for Strategic Affairs), I could not have imagined the scope of the work and the openings it represented.

Knowing, understanding, deciding: this was the triptych that structured the strategic approach at that time. Today, we are at the stage of "knowledge and anticipation" and the emergence of a new strategic function. Of course, necessity is the law and there is an urgent need to give ourselves the means to understand the evolution of the strategic context; but in order to get there, the political authorities have to admit the possibility or existence of such a capacity.

The initial credit for this goes to Jacques Arcade, whose work, reflections, advice, and innovations have made it possible to conceptualize an efficient working method and to conduct studies of great relevance within the Ministry of Defense.

This book is therefore timely, coming at a time when, in France, the tools of defense – in the broadest sense of the term – are evolving. It presents a major innovation in the way to approach the identification of future challenges, the definition of fundamental objectives, and the implementation of an effects-based strategy. This is useful in the field of defense, of course, but also – as is demonstrated throughout the book – in all areas where decision-making is of a strategic nature.

For the skeptics of anticipation, it also demonstrates that prospective analysis and strategic engineering are part of an approach integrating rigorous methods and scientific foundations. This is a far cry from the "wet-finger" and "corner-of-the-table" approaches that some people like to imagine. The tools to assist in decision-making do exist. They are credible, and in many cases, have been proven. They are not "black boxes", but rather instruments that provide aid and assistance, returning only what humanity, the center of prospective reflection, has been willing to enter.

The primary interest of strategic engineering, as Jacques Arcade explains, lies both in the ability to solve complex problems by decomposition and recomposition into fractals (the nested frames), as well as – and above all – in the creation of a steering system or device for decision-making, the implementation of a strategy and the management of action with a clear appreciation of the room for maneuvering, the risks, and the performance. Strategic engineering provides an instrument for steering by effects and enables proactivity, that is, the preparation of a change that is anticipated or even provoked. This places us at the very heart of strategy.

The demands of the era of sustainable development have led to a renewed interest in foresight, which is to say, the study of possible futures. In addition to the openness it brings as a remedy for the prevailing catastrophism, foresight frees us from fatalism, as Gaston Berger said. Strategic engineering, in the same way, provides hope in the capacity to take control of the currents generated by globalization and to prepare a credible strategy. It is helpful to clarify uncertainties, which is in itself a first step to reassuring decision-makers. It obliges former experts to give their opinions, to evaluate and weight the elements of appreciation, which makes it possible to objectively estimate the room for maneuvering. It takes full account of the human factor by providing criteria for discerning the roles and strategies of each of the players – and this is a very rich innovation. It provides a clear and operational vision of what we know, what we have mastered, and, conversely, what we need to deepen, research, and study.

Strategic engineering is therefore a method, and in addition, a guide for enlightened action.

This book is very rich. With the hindsight of several studies carried out at the Délégation aux affaires stratégiques and several strategic engineering seminars conducted for the Collège interarmées de Défense and the École de Guerre, I can testify that the application of the theories, methods, and tools described in this book has produced convincing results. Since 2013, as the head of a master's degree program at the Institut d'études politiques de Rennes in France, where these methods are taught, I can attest to the interest aroused among students and their effectiveness in successfully dealing with complex subjects in a limited time frame.

Like any innovative work, putting it into practice is a daring feat. This is not a book of recommendations for future gray eminences. It is a guide for all those who prepare high-level decisions, for those whose ambition is to empower authority with decision-making abilities at the strategic or political level, to see far in the future, to understand, and to anticipate in order to make the right decisions at the right time, despite "adversity".

It is the hope of any manager to be able to be pre-active, and better yet, proactive. As Seneca said: "If one does not know to which port one is sailing, no wind is favorable."

I would therefore like to thank Jacques Arcade for what his work has already made possible, and even more for the successes to come.

Philippe BOONE
General (2S)
Formerly Deputy Director for International Relations at the
Collège interarmées de défense, French Ministry of Defense

Co-director of the master's degree
in Security, Defense and Strategic Intelligence
(SE-DEFIS) at the Institut d'études politiques, Rennes, France

Glossary

Acting entities: natural or legal persons, organizations, institutions, or nation-states maintaining relations with the actors on the sociopolitical stage, and capable of taking initiative as an actor.

Anticipation: the act of taking a long-distance look into the future, transforming this view into an incentive to act and defining one's own objectives.

Ascending actors: the power exercised by one of the actors through its relationship (direct and/or indirect) with another actor. This ascendancy can be tangible and of great size (Goliath) and/or subtle and postured (David). The product of the ascendants of posture and stature is equal to the balance of power between the actors.

Attribute scale: a qualitative scoring system; a scoring grid, a qualitative appreciation of a given system.

Autopoiesis: a process that allows a living system to organize itself, to regenerate itself while preserving its own identity.

Background: the background of a given situation, grid, or generic frame on which particular elements are drawn.

Boundary marker: the terminal (second member) of a mathematical constraint considered as a parameter that can be varied.

Brother criteria: a set of criteria descending directly from the same node of the tree, considered as a "father".

Cascading weighting: the relative weight assigned to a macrocriterion (the root of the tree) obtained from a group of local weights, assigned to the different nodes of the tree.

Ceiling/floor weight: the maximum or minimum relative share that an action can represent in a composition, taking into account the room for maneuver.

Chain of incidence: a succession of consequences that spread to different levels.

Challenge power: the ability to trigger mobilization and reaction on the part of decision-makers.

Circumstantial assessment: a qualitative assessment taking into account the circumstances and context.

Combination of hypotheses: a set of combinations of hypotheses that are relevant, consistent, and plausible.

Complexity: a state inherent in an uncertain and destabilizing situation; it is reflected in the multiplicity of elements and their relationships.

Component tree: a hierarchy of elements and parts of a set in nested frames. This corresponds to a diagonal section of the elements at different levels or scales. The tree does not take into account the interactions of elements at different levels.

Core capability: the core business or key ability of an organization or company, which is distinctive, relatively inimitable, and is expected to play a pivotal role in its future evolution.

Critical circumstances: conditions or elements that require an analysis and re-examination of the situation.

Critical condition: a condition that if not fulfilled may jeopardize the unwinding and/or the expected results.

Context: all the conditions, circumstances, and "ins and outs" of a given situation.

Contrasting images: the typical representations of the future, clearly differentiated and relatively exclusive from each other.

Contrasting scenarios: the scenarios that result from profiles of assumptions about the future with regard to factors of uncertainty considered major within the prospective horizon. These assumptions are mutually exclusive or even incompatible.

The scenarios produced by such assumptions are all the more contrasting as their assumptions are distinct.

Corpus of constraints: a set of interdependent constraints.

Critical node: the node in a tree of criteria whose weighting can change the priorities and results of the intervention with a small variation in weighting.

Dashboard: simultaneous visualization (Simoptis tool) of the possibilities space and performance and action profiles.

Decision: the ability to decide between alternatives in order to serve one's best interests.

Decision-making situation: the context in which the decision-making process takes place. Its formalization can result in its inclusion on a political agenda.

Degree of criticality: the extent to which the critical nature of a factor, a variable, or a point is felt and imposes itself on the analyst.

Domain of possibility: a visual representation of the system of constraints and the space of possibilities that it induces.

Encompassing frame: a container or frame that contains lower level frames at a finer scale, reflecting the conditions and circumstances of the frames at the finer level.

Evaluation system: a set of criteria, weightings, and methods for assessing given objects.

Exploratory scenario: a scenario that deals with the uncontrollable dimensions of the future, as opposed to normative scenarios or strategic options that are subject to the actor's will.

Field of investigation: all the relevant fields or dimensions covered by a given discipline, method, or approach.

Flesh: the way in which auxiliary hypotheses relating to the various scenarios are oriented.

Founding hypothesis: a hypothesis around which the creation of scenarios is articulated and which contributes to the determination of their structure.

Fractal: a sequence of objects at different scales that have certain similarities. There is a form of inter-scale invariance of adventitiously shaped extension of form and context.

Framework: the morphological profiles of scenarios based on structural assumptions about the future as answers to key questions about major uncertainties.

Fundamental goal: the intention of an action taken and what is used to justify it.

General context of intervention: a specific description of the sociopolitical actors' interplay and classical dimensions (description of economic and legal situations, regulations, environment). This represents a relatively immediate view of the situation.

General environment scenario: an exploratory scenario that focuses on the context and not on the interventions of the actor concerned.

Geometric envelope of solution profiles: a visualization of the smallest space, including all possible and conceivable solution profiles.

Global systemic framework: a broad system of representation of a situation that takes into account interactions at different levels and in the broadest possible way, and covers the whole of the system.

Glocal: a characterization of an object that is found to be both local and global.

Hold: the influence and control of a system (actors' interplay) over another system (intervention).

Hypertext: a process by which it is possible to gain more depth on a particular element in order to contextualize it with the possibility of colliding, repeating this type of deepening at different levels.

Impact: the direct or indirect influence or consequences of an action in its context.

Incidence: the direct or indirect consequence of a given action or event.

Incompatibility threshold: the level at which constraints become incompatible and the problem no longer allows for a solution.

Intentional requirements: a requirement set voluntarily, including to oneself.

Intervention: a structuring action most often carried out in a delicate, critical, or crisis situation.

Invariant: a relatively predetermined element on the horizon of foresight, such as the ageing of the French population, for example; a property or characteristic that is constant over time or redundant in a given element.

Key question: the transcription of a major uncertainty and/or strategic lever in a way that is "engaging" to the analyst or decision-maker, in other words, one that calls for a range of possible answers.

Law of composition: the logic of combining and mixing elements.

Lever: an elementary means of action that is taken and which offers a range of possibilities or options; control element that causes consequences whose position affects the system.

Macrocriterion: the root of a criteria "tree"; it is the trunk of a criteria tree.

Major trends: elements which characterize the present situation, whose evolution can be considered relatively continuous and predetermined.

Major uncertainty: a relatively unpredictable and relatively unavoidable element that we do not control and that opens up the field of possibilities.

Matrix of actors' interplay: a square table with two entries (actor/actor) the diagonal boxes of which are the signage (identity card, objectives, projects, preoccupations, interests) and the non-diagonal boxes of which contain the actors' cross-influences on each other.

Metasystem: a system of systems, that is, a complex set of interrelated systems; at a higher level, this also refers to the system of encompassing functions of an operating system.

Morphological analysis: analysis of the structure of an element in the form of combinations or contrasting profiles of elementary constituents.

Morphological profile: a contrasting profile linking different hypotheses respectively associated with uncertainties about the future.

Multicriteria analysis: a decision/support method, allowing for a comparative assessment between different projects or options on the basis of different criteria, in this case scenarios, weighted respectively by their probability of occurrence.

Multicriteria evaluation: a comparative evaluation method based on the rating of competing projects from different weighted perspectives.

Nested components: elements nested within each other with more than one element per level.

Nested frames: a series of frames in the form of “Russian dolls”; a visual or figurative mode of representation that refers to the notion of context. Each frame is immersed in a larger frame that reflects the context (i.e. its conditions and circumstances).

Optional intervention profile: a combination (coherent, relevant, and credible) of basic options (or modes of action) respectively recommended according to different strategic levers.

Organization: an institutionalized entity in the sense of a living system, possibly made up of human elements. In the theory of living systems, the space of possible combinations between different elements. The structure of the organization corresponds only to a particular combination of elements.

Orientation: the direction of the tendency of the action.

Overall performance: the result of the evaluation of an intervention in relation to the overall systemic framework and therefore to all the macrocriteria it contains.

Overriding goal: an ambition, goal, or objective the satisfaction of which is of an insatiable nature.

Performance window (or range): range of performance levels that allows one to trace the room for maneuver.

Pivotal strategy: a strategy that is robust to circumstances and can be mixed with other diverse strategies to deal with a variety of situations.

Planning: a method to implement and organize the means necessary for a directed action.

Prerequisite: a necessary precondition for the implementation of a given process.

Preventive locking: a careful posture to limit damage whatever the future context or situation.

Prioritization: the consideration of the priorities ordered in a tree-like manner and/or in terms of possible weighting; that is, according to their relative weights.

Proactivity: the ability to push the limits of the maneuvering space to one's advantage in order to surpass oneself. It allows one to open up new horizons, overcome obstacles, reconfigure oneself, and/or transcend a situation.

Probability of occurrence: the probability that an event, scenario, or situation will occur.

Prohibitive level of vulnerability: a level of risk beyond which the integrity and/or validity of the object concerned is at risk or the validity of the objective concerned is questioned.

Prospective/foresight: an essentially exploratory method of anticipation that applies more to the context and the environment, that is, the future as a whole, than to the particular action likely to be developed there.

Purpose: a comprehensive and far-reaching goal of an action or intervention that justifies the rationale for the intervention.

Redundancy threshold: the level at which certain constraints become useless, obsolete, or inactive; these are automatically confirmed by the effective respect of other constraints to which the system is subjected.

Room for maneuver: the space available to freely carry out actions.

Salient contradiction: a contradiction significant enough to be highlighted.

Salient point: a singular point, that is, one that stands out relative to a more uniform background. It represents a critical point in the sense that it is decisive in understanding the essence of a situation or in achieving a result from a situation. The evolution of a situation is very sensitive to the way in which a salient point is approached.

Sapience: the fruit of a subtle connection or a clever combination of science and wisdom.

Self-similarity: a property according to which the structure of a whole is similar to that of its parts, whatever the scales of observation. This geometrical property called "scale invariance" is characteristic of "fractal" objects, which are very common in natural and in the life sciences.

Semantic summary: title translating the content of a profile or scenario or option.

Spiral: a geometrical representation of a recursive phenomenon.

Strategic engineering: the art of conceiving a decision along the way, promoting the overcoming of limits and margins or room for maneuver. This discipline takes into account the actors, the balance of power, and the context by removing obstacles to ensure the viability of the system.

Strategic intervention domain: a relevant and semantic grouping of strategic directions.

Strategic landscape: a domain broader than the context, this includes all the elements of the context by adding the notions of the interests of the actors' interplay as well as the elements of construction and assets that create inertia in the action that one carries out. This refers to the environment in which the action must be immersed.

Strategic lever: the inescapable strategic action point on which one has control and which opens up the range of actions.

Strategic option: the optional profile of a particular intervention.

Strategic posture: the logic of the decisions we make about the future, based on the nature of the uncertainty.

Strategic referential: a concept borrowed from Hamel and Prahalad (1999) that expresses the magnetic field that governs the action and understanding of the situation; this includes the driving force, vision, belief, culture, purpose, and/or fundamental capabilities, as well as bold objectives.

Strategy: a sequence of organization of means to remove obstacles to achieve one's ambition. In the framework of strategic engineering, strategy is the creation of the viability of the system by articulating the ambition and the goal with the evolution of the action along the way. It is a proactive approach to strategy.

Structural analysis: a method for deciphering the structure and trend dynamics of a system of factors or variables by identifying their interactions in a table and then projecting them visually onto a graph showing influence against dependency.

Structural coupling relationship: a stable interdependence between processes and/or the elements from which they are formed.

Subsystem: a part of a system which itself can be considered as a system.

System: a complex set of interrelated elements that at least partially obey a specific logic of operation or purpose.

Systemic complement: starting from a hierarchy of components, this consists of completing this hierarchy by mapping the components at each level.

Threshold setting: the determination of a minimum level for a given quantity.

Tree structure: organization of increasingly fine elements in the form of a tree (trunk, branch, leaf).

Zone of viability: a space within which a unit (or system), whether living or not, asserts its identity by differentiating itself and attempts to ensure its sustainability over a period of time relevant to its purpose.

References

Arcade, J. (1990). La dynamique de l'échange composite : une interprétation analytique. *Revue d'économie appliquée*, LXIII(4).

Arcade, J. (1995). Approche systémique de l'action stratégique. Course, Institut d'études économiques, sociales et techniques des organisations (IESTO/CNAM), Paris, Montpellier.

Arcade, J. (1998). *Articuler prospective et stratégie : parcours du stratège dans la complexité*. Futuribles International, Paris.

Arcade, J. (1999). Conjuguer rigueur et pragmatisme en prospective stratégique. In *La R&D prospective-essai*. Réseau Intelligence de la Complexité, Association européenne du programme Modélisation de la Complexité (MCX-APC), Aix-en-Provence.

Arcade, J. (2000). *Mettre en scène des futurs inédits : éléments d'ingénierie prospective à l'usage des praticiens*. Futuribles International, Paris.

Arcade, J. (2004a). Business Angels et secteurs émergents : entre perspicacité et proactivité. *4th Monte Carlo Business Angels Forum*, Monte Carlo.

Arcade, J. (2004b). L'économie sociale prend de la hauteur pour se dépasser. *Les Rencontres du Mont-Blanc, Forum international des dirigeants de l'économie sociale et solidaire*, Morzine.

Arcade, J. (2009). L'intégration de services : une exigence d'innovation organisationnelle. *Le blog du planneur stratégique*, Paris.

Arcade, J. (2017). Comprendre les catégories de risques et d'environnements pour s'affranchir des erreurs d'ordre stratégique. *La Lettre de XMP Consult*, 3.

Arcade, J. and Gille, L. (1996). La prospective des télécoms. In *Conférences du Conservatoire 1994–1995, La Mémoire du XXème siècle sur la prospective, l'innovation et les territoires*. National Conservatory of Arts and Crafts, Paris.

Arcade, J. and Misonne-Lemaire, C. (1995). Redefining the core competencies: Simulating/Optimizing with the Simopt System. In *Scenarios Building: Convergence and Differences, Proceedings of Profutures Workshop*. Institute for Prospective Technological Studies, European Commission, Joint Research Center.

Arcade, J. and Perrin, F. (2001). Articulation prospective, stratégie, veille : vers une nouvelle approche du risque stratégique. *Conférence AFPLANE.*

Ashby, W.R. (1963). *Principles of the Self-organizing System*. Pergamon Press, Oxford.

Attali, J. (2006). *Une brève histoire de l'avenir*. Fayard, Paris.

Avenier, M.J. (1992). *Pilotage stratégique de l'entreprise, éco-système complexe*. Atelier, GRASCE, Aix-en-Provence.

Barker, J.A. (1995). *Les paradigmes : à la découverte du futur*. Un monde différent, Quebec.

Baumard, P. (1996). *Prospective à l'usage des managers*. Litec, Paris.

Benassouli, P. and Monti, R. (1995). La planification par scénarios : le cas Axa France 2005. *Revue Futuribles*, 203.

Benkirane, R. (2002). *La complexité, vertiges et promesses : 18 histoires de sciences*. Le Pommier, Paris.

Berger, G. (1964). *Phénoménologie du temps et prospective*. PUF, Paris.

Berthoz, A. (2009). *La simplexité*. Odile Jacob, Paris.

Berthoz, A. (2013). *Le cerveau créateur de mondes*. Odile Jacob, Paris.

Blanchard, K. and Waghorn, T. (1997). *Anticiper le changement*. Dunod, Paris.

Boyer, L. and Equilbay, N. (1986). *Le projet d'entreprise*. Éditions d'Organisation, Paris.

Cadet, B., Grenier, C., Smida, A. (2003). *Les décisions sous contraintes : une approche pluridisciplinaire*. Presses universitaires de Caen, Caen.

Calori, R. and Atanier, T. (1989). *L'action stratégique : le management transformateur*. Éditions d'Organisation, Paris.

Cassé, M. (2014). *Du vide et de l'éternité*. Odile Jacob, Paris.

Caune, A., Lichnerowicz, A., Schützenberger, M.P. (2000). *Triangle de pensées*. Odile Jacob, Paris.

Changeux, J.P. and Caune, A. (2000). *Matière à penser*. Odile Jacob, Paris.

Cohen, B., Shenk, J., Gore, A. (2017). *Une suite qui dérange : le temps de l'action*. Documentary film, Paramount Pictures, Los Angeles.

Collins, J. (2013). *De la performance à l'excellence : devenir l'entreprise leader*. Pearson, Montreuil.

Comte-Sponville, A. and Ferry, L. (1999). *La sagesse des modernes*. Robert Laffont, Paris.

Conseil national de prospective et de planification stratégique (2005). Étude nationale prospective "Burkina 2025". General report.

Cova, B. and Wickham, S. (1996). *Stratégies d'incertitude*. Economica, Paris.

Crozier, M. and Friedberg, E. (1997). *L'acteur et le système*. Le Seuil, Paris.

Dantzig, G.B. (1966). *Applications et prolongements de la programmation linéaire*. Dunod, Paris.

Dathané, S. (2015). *Décider dans un monde complexe*. Maxima, Paris.

De Jouvenel, B. (1968). *Arcadie : essai sur le mieux vivre*. Futuribles, Paris.

De Jouvenel, B. (1976). *La civilisation de puissance*. Fayard, Paris.

De Jouvenel, H. and Delcroix, G. (2008). *Le choc des identités ? Cultures, civilisations et conflits de demain*. Futuribles, Paris.

De Rosnay, J. (1995). *L'homme symbiotique : regards sur le troisième millénaire*. Le Seuil, Paris.

Declerck, R.P., Eymeri, P., Créner, M.A. (1980). *Le management stratégique des projets*. Hommes et Techniques, Paris.

Desportes, V. and Phelizon, B. (2007). *Introduction à la stratégie*. Economica, Paris.

Drew, M. and William, R. (2012). *Pendulum: How Past Generations Shape Our Present and Predict Our Future*. Vangar Press, New York.

Dugué, B. (2014). *Le sacre du vivant, essai post-darwinien sur l'essence de la vie et son évolution*. Temps Présent Éditions, Paris.

Dupuy, J.P. (2004). *Pour un catastrophisme éclairé*. Le Seuil, Paris.

Easton, D. (1965). *A System Analysis of Political Life*. John Wiley and Sons, Hoboken, New Jersey.

ECOSIP (1993). *Pilotages de projet et entreprises*. Economica, Paris.

ECOSIP (1996). *Cohérence, pertinence et évaluation*. Economica, Paris.

Ewald, F., Gollier, C., de Sadeleer, N. (2001). *Le principe de précaution*. PUF, Paris.

Falque, L. and Bougon, B. (2013). *Pratiques de la décision : développer ses capacités de discernement*. Eyrolles, Paris.

Fauvet, J.-C. (2004). *L'élan sociodynamique*. Éditions d'Organisation, Paris.

Figari, G. (1994). *Évaluer : quel référentiel ?* De Boeck Supérieur, Louvain-la-Neuve.

Garcia, E.J. (2013). *Le génie du leadership*. Dunod, Paris.

Génelot, D. (1992). *Manager dans la complexité : réflexions à l'usage des dirigeants*. INSEP Éditions, Paris.

Gleick, J. (2008). *La théorie du chaos*. Flammarion, Paris.

Godet, M. (1991). *De l'anticipation à l'action*. Dunod, Paris.

Godet, M. (1997a). *Manuel de prospective stratégique 1*. Dunod, Paris.

Godet, M. (2007b). *Manuel de prospective stratégique 2*. Dunod, Paris.

Godet, M. and Arcade, J. (1977). Critique des termes de l'échange et conséquences pour le Tiers Monde. *Revue Tiers Monde*, XVIII(70), April–June.

Gomes-Casseres, B. (2015). *Remix Strategy: The Three Laws of Business Combinations*. Harvard Business Review Press, Brighton.

Gonod, P. (1996). *Dynamique des systèmes et méthodes prospectives*. Futuribles, Paris.

Gore, A. (2019). *De nouvelles idées pour la crise climatique*. TED Talks, New York.

Grouard, B. and Meston, F. (2005). *L'entreprise en mouvement : conduire et réussir le changement*. Dunod, Paris.

Guigou, J.L. (2006). *Réhabiliter l'avenir : la France malade de son manque de prospective*. L'Harmattan, Paris.

Guigou, J.L. and DATAR (2002). *Aménager la France de 2020 : mettre les territoires en mouvement*. La Documentation Française, Paris.

Guillen, M. (1992). *Des ponts vers l'infini. Des mathématiques à figure humaine*. Albin Michel, Paris.

Hamel, G. (2008). *La fin du management, inventer les règles de demain*. Vuibert, Paris.

Hamel, G. and Prahalad, D. (1999). *La conquête du futur : construire l'avenir de son entreprise plutôt que le subir*. Dunod, Paris.

Heesen, B. (2016). *Effective Strategy Execution*. Springer, Berlin/Heidelberg.

Helmer, O. (1983). *Looking Forward: A Guide to Future Research*. SAGE Publications, Thousand Oaks.

Herbemont, P. and Cesar, B. (1996). *La stratégie du projet latéral*. Dunod, Paris.

Hipel, K.W. and Fraser, N.M. (1991). Cooperation in conflict analysis. *Cahier du Lamsade*, 103.

Husserl, E. (1985). *Idées directrices pour une phénoménologie*. Gallimard, Paris.

IDATE, Proactivité Conseil and Cret-Log (2005). Le commerce électronique : quel impact à long terme sur le territoire francilien ? Report, Ministère de l'Équipement, des Transports et du Logement.

Jain, P. and Sharma, P. (2014). *Behind Every Good Decision*. McGraw-Hill Education, New York.

Jaworsky, J. (2011). *Synchronicity, the Inner Path of Leadership*. Berret-Koehler Publishers, Oakland.

Kaplan, R.S. (2015). *What Do You Really Need to Lead: The Power of Thinking Like an Owner*. Harvard Business Review Press, Brighton.

Kauffman, S.A. (2013). *Réinventer le sacré, une nouvelle vision de la science, de la raison et de la religion.* Dervy, Paris.

Kim, W.C. and Mauborgne, R. (2010). *Stratégie Océan Bleu : comment créer de nouveaux espaces stratégiques*. Pearson, Montreuil.

Kourilsky, P. (2001). *Du bon usage du principe de précaution.* Odile Jacob, Paris.

Labruffe, A. (2013). *Maîtriser le changement.* Afnor Éditions, La Plaine Saint-Denis.

Lapierre, J.-W. (1968). *Essai sur le fondement du pouvoir politique*. Ophrys, Gap.

Le Boterf, G. (1994). *De la compétence*. Éditions d'Organisation, Paris.

Le Moigne, J.L. (1990). *La modélisation des systèmes complexes*. Dunod, Paris.

Leibniz, G.W. (2013). *La monadologie : édition intégrale.* Ink Book, Paris.

Lerbet, G. (1993). *Approche systémique et production de savoir*. L'Harmattan, Paris.

Lesourne, J. and Stoffaës, C. (1996). *La prospective stratégique d'entreprise.* InterÉditions, Paris.

Lorino, P. (1995). *Comptes et récits de performance : essai sur le pilotage de l'entreprise*. Éditions d'Organisation, Paris.

Lynch, D. and Kordis, P. (1994). *La stratégie du Dauphin : les idées gagnantes du XXI*[ème] *siècle.* Les Éditions de l'Homme, Montreal.

Mandelbrot, B. (1982). *Fractal Geometry of Nature*. W.H. Freeman and Company, New York.

Mandelbrot, B. (2009). *Une approche fractale des marchés*. Odile Jacob, Paris.

Mandelbrot, B. (2010). *Les objets fractals : forme, hasard et dimension.* Flammarion, Paris.

Mangham, C., McGrath, P., Reid, G., Stewart, M. (1995). Résilience, pertinence dans le contexte de la promotion de la santé. Working paper, Atlantic Health Promotion Research Centre, Dalhousie University, Halifax.

March, J.G. and Simon, H. (1969). *Les organisations*. Dunod, Paris.

Martinet, A.C. (1990). *Épistémologies et sciences de gestion.* Economica, Paris.

Massé, P. (1965). *Le plan ou l'anti-hasard.* Gallimard, Paris.

Merleau-Ponty, M. (1976). *Phénoménologie de la perception.* Gallimard, Paris.

Minsky, D. (1997). *La société de l'esprit.* InterÉditions, Paris.

Mintzberg, H. (1994). *Grandeur et décadence de la planification stratégique*. Dunod, Paris.

Mintzberg, H. (2003). *Le pouvoir dans les organisations*. Eyrolles, Paris.

Mintzberg, H., Ahlstrand, B., Lampel, J. (1999). *Safari en pays stratégie : l'exploration des grands courants de la pensée stratégique*. Pearson, Montreuil.

Miquel, P.A. (2000). *Comment penser le désordre ?* Fayard, Paris.

Moles, A.A. (1990). *Les sciences de l'imprécis*. Le Seuil, Paris.

Moreau, F. (2002). *Comprendre et gérer les risques*. Éditions d'Organisation, Paris.

Morin, E. (2008). *La Méthode en 6 volumes*. Le Seuil, Paris.

Muller, P. (2013). *Les politiques publiques*. PUF, Paris.

Musso, P. (1996). *Communiquer demain : nouvelles technologies de l'information et de la communication*. Éditions de l'Aube, Paris.

Musso, P. (ed.) (2008). *Territoires et cyberespaces en 2030*. La Documentation Française, DIACT, Paris.

Musso, P. (2014). *L'imaginaire industriel*. Manucius, Paris.

Musso, P., Ponthou, L., Seuillet, E. (2005). *Fabriquer le futur : l'imaginaire au service de l'innovation*. Pearson, Montreuil.

Napoli, B. (2001). L'expérience du FOREM en matière de prospective. In *Évaluation, prospective et développement régional*, Destatte, P. (ed.). Institut Jules-Destrée, Charleroi.

Nash, J.F. (1950a). *Equilibrium points in N-Person Games*. Princeton University Press, Princeton.

Nash, J.F. (1950b). The bargaining problem. *Econometrica*, 18(2), 155–162.

von Neumann, J. and Morgenstern, O. (1953). *Theory of Games and Economic Behavior*. Princeton University Press, Princeton.

Normann, R. (1994). *Le management des services*. InterÉditions, Paris.

Nottale, L. (1998). *La relativité dans tous ses états*. Hachette, Paris.

Pelt, J.M. and Rabhi, P. (2014). *Le monde a-t-il un sens ?* Fayard, Paris.

Perroux, F. (1973). *Pouvoir et économie*. Dunod, Paris.

Perroux, F. (1981). *Pour une philosophie du nouveau développement*. Les Presses de l'Unesco, Paris.

Piaget, J. (1957). *Logique et équilibre*. PUF, Paris.

Piaget, J. (1967). *Biologie et connaissance : essai sur les relations entre les régulations organiques et les processus cognitifs*. Gallimard, Paris.

Piaget, J. (1971). *La psychologie de l'intelligence*. Armand Colin, Paris.

Piaget, J. (1975). *L'équilibration des structures cognitives : problème central du développement*. PUF, Paris.

Porter, M. (1986). *Choix stratégiques et concurrence*. Economica, Paris.

Porter, M., Collins, J., Christensen, C., Kim, W.C., Mauborgne, R., Neilson, G., Martin, C. (2015). *Le must de la stratégie : théories majeures pour mener votre entreprise vers le succès*. Harvard Business Review France, Paris.

Portnoff, A.-Y. (2003). *Sentiers d'innovation*. Futuribles International, Paris.

Prigogine, I. (1996). *La fin des certitudes : temps, chaos et lois de la nature*. Odile Jacob, Paris.

Prigogine, I. (1999). *De l'être au devenir (entretiens)*. Stanké Éditions, Liège.

Probst, G.J.B. and Buchel, B.S.T. (1995). *L'Entreprise apprenante*. Éditions d'Organisation, Paris.

Probst, G.J.B., Mercier, J.Y., Bruggimann, O., Rakotobarrison, A. (1992). *Guider le développement de l'entreprise*. Éditions d'Organisation, Paris.

Rabhi, P. (2010). *Vers la sobriété heureuse*. Actes Sud, Arles.

Ribordy, L. (2007). *Nombre et géométrie : vers un peu de lumière*. Maison de Vie, Paris.

Ricard, M. and Thuan, T.X. (2000). *L'infini dans la paume de la main : du Big Bang à l'éveil*. Fayard, Nil Éditions, Paris.

Richardson, L.F. (2007). *Weather Prediction by Numerical Process*. Cambridge University Press, Cambridge.

Rifkin, J. (2000). *L'âge de l'accès : la révolution de la nouvelle économie*. La Découverte, Paris.

Rifkin, J. (2012). *La Troisième révolution industrielle : comment le pouvoir latéral va transformer l'énergie, l'économie et le monde*. Les liens qui libèrent, Paris.

Rifkin, J. (2014). *La nouvelle société du coût marginal zéro*. Les liens qui libèrent, Paris.

Roddier, F. (2012). *Thermodynamique de l'évolution : un essai de thermo-bio-sociologie*. Parole, Artignosc-sur-Verdon.

Roubelat, F. (1996). La prospective stratégique en perspective : genèse, étude de cas, prospective. PhD thesis, National Conservatory of Arts and Crafts, Paris.

Roy, B. (1970). *Algèbre moderne et théorie des graphes tomes 1 et 2*. Dunod, Paris.

Roy, B. (1993). *Aide multicritère à la décision*. Economica, Paris

Saaty, T.L. (1990). *The Analytic Hierarchy Process: Multicriteria Decision Making*. RWS Publications, Pittsburgh.

Saint-Paul, R. and Tenière-Buchot, P.F. (1974). *Innovation et évaluation technologiques : sélection des projets, méthodes de prévision*. Entreprise moderne d'édition, Paris.

Sapoval, B. (1997). *Universalités et fractales : jeux d'enfants ou délits d'initié ?* Flammarion, Paris.

Savall, H. and Zardet, V. (2005). *L'ingénierie stratégique du roseau*. Economica, Paris.

Scharling, A. (1985). *Décider sur plusieurs critères*. Presses polytechniques et universitaires romandes, Lausanne.

Serres, M. (2015). *Le gaucher boîteux : puissance de la pensée*. Le Pommier, Paris.

Sfez, L. (1992). *Critique de la décision.* Presses de Sciences Po, Paris.

Shiba, S. (1995). *Le management par percée : méthode HOSCHIN.* INSEP Éditions, Paris.

Simon, H.A. (1991). *Sciences des systèmes, sciences de l'artificiel.* Dunod, Paris.

Staune, J. (2015). *Les clés du futur : réinventer ensemble la société, l'économie et la science.* Plon, Paris.

Stowell, S.J. and Mead, S.S. (2016). *The Art of Strategic Leadership.* John Wiley and Sons, Hoboken, New Jersey.

Thom, R. (1977). *Stabilité structurelle et morphogénèse.* InterÉditions, Paris.

Thom, R. (1993). *Prédire n'est pas expliquer.* Flammarion, Paris.

Tixier, N. and Faucheux, E. (eds) (2007). L'architecture de la grande échelle : vers la simulation d'éco-systèmes urbains. Interdisciplinary research program PUCA 2006–2007, Ministère de la Culture.

Tonnelé, A. (2014). *65 outils pour accompagner le changement individuel et collectif.* Éditions d'Organisation, Paris.

Varela, F. (1989a). *Autonomie et connaissance : essai sur le vivant.* Le Seuil, Paris.

Varela, F. (1989b). *Connaître. Les sciences cognitives, tendances et perspectives.* Le Seuil, Paris.

Varela, F. (1994). *L'arbre de la connaissance.* Addison-Wesley, Boston.

Varela, F. (2017). *Le cercle créateur : écrits (1976–2001).* Le Seuil, Paris.

Vernette, P. (1987). Identifier les attributs déterminants : une comparaison de six méthodes. *Recherche et applications en marketing*, 2(4), 1–21.

Vincke, P. (1989). *L'aide multicritère à la décision.* Ellipses, Paris.

Vivier, L. (2004). *La topologie : l'infini maîtrisé.* Le Pommier, Paris.

Wanty, J. and Federwish, J. (1969). *Modèles globaux d'économie d'entreprise.* Dunod, Paris.

Wilson, E.O. (2000). *L'unicité du savoir : de la biologie à l'art, une même connaissance.* Robert Laffont, Paris.

Wuffli, R.A. (2015). *Inclusive Leadership: A Framework for the Global Era.* Springer, Berlin.

Zwicky, F. (1969). *Discovery, Invention, Research: Through the Morphological Approach.* Macmillan Publishers, New York.

Index

T

Other titles from

ISTE

in

Innovation, Entrepreneurship and Management

2021

BOBILLIER CHAUMON Marc-Eric
Digital Transformations in the Challenge of Activity and Work: Understanding and Supporting Technological Changes
(Technological Changes and Human Resources Set – Volume 3)

2020

ACH Yves-Alain, RMADI-SAÏD Sandra
Financial Information and Brand Value: Reflections, Challenges and Limitations

ANDREOSSO-O'CALLAGHAN Bernadette, DZEVER Sam, JAUSSAUD Jacques, TAYLOR Robert
Sustainable Development and Energy Transition in Europe and Asia
(Innovation and Technology Set – Volume 9)

BEN SLIMANE Sonia, M'HENNI Hatem
Entrepreneurship and Development: Realities and Future Prospects
(Smart Innovation Set – Volume 30)

CHOUTEAU Marianne, FOREST Joëlle, NGUYEN Céline
Innovation for Society: The P.S.I. Approach
(Smart Innovation Set – Volume 28)

CORON Clotilde
Quantifying Human Resources: Uses and Analysis
(Technological Changes and Human Resources Set – Volume 2)

CORON Clotilde, GILBERT Patrick
Technological Change
(Technological Changes and Human Resources Set – Volume 1)

CERDIN Jean-Luc, PERETTI Jean-Marie
The Success of Apprenticeships: Views of Stakeholders on Training and Learning
(Human Resources Management Set – Volume 3)

DELCHET-COCHET Karen
Circular Economy: From Waste Reduction to Value Creation
(Economic Growth Set – Volume 2)

DIDAY Edwin, GUAN Rong, SAPORTA Gilbert, WANG Huiwen
Advances in Data Science
(Big Data, Artificial Intelligence and Data Analysis Set – Volume 4)

DOS SANTOS PAULINO Victor
Innovation Trends in the Space Industry
(Smart Innovation Set – Volume 25)

GASMI Nacer
Corporate Innovation Strategies: Corporate Social Responsibility and Shared Value Creation
(Smart Innovation Set – Volume 33)

GOGLIN Christian
Emotions and Values in Equity Crowdfunding Investment Choices 1: Transdisciplinary Theoretical Approach

GUILHON Bernard
Venture Capital and the Financing of Innovation
(Innovation Between Risk and Reward Set – Volume 6)

LATOUCHE Pascal
Open Innovation: Human Set-up
(Innovation and Technology Set – Volume 10)

LIMA Marcos
Entrepreneurship and Innovation Education: Frameworks and Tools
(Smart Innovation Set – Volume 32)

MACHADO Carolina, DAVIM J. Paulo
Sustainable Management for Managers and Engineers

MAKRIDES Andreas, KARAGRIGORIOU Alex, SKIADAS Christos H.
Data Analysis and Applications 3: Computational, Classification, Financial, Statistical and Stochastic Methods
(Big Data, Artificial Intelligence and Data Analysis Set – Volume 5)
Data Analysis and Applications 4: Financial Data Analysis and Methods
(Big Data, Artificial Intelligence and Data Analysis Set – Volume 6)

MASSOTTE Pierre, CORSI Patrick
Complex Decision-Making in Economy and Finance

MEUNIER François-Xavier
Dual Innovation Systems: Concepts, Tools and Methods
(Smart Innovation Set – Volume 31)

MICHAUD Thomas
Science Fiction and Innovation Design (Innovation in Engineering and Technology Set – Volume 6)

MONINO Jean-Louis
Data Control: Major Challenge for the Digital Society
(Smart Innovation Set – Volume 29)

MORLAT Clément
Sustainable Productive System: Eco-development versus Sustainable Development
(Smart Innovation Set – Volume 26)

SAULAIS Pierre, ERMINE Jean-Louis
Knowledge Management in Innovative Companies 2: Understanding and Deploying a KM Plan within a Learning Organization
(Smart Innovation Set – Volume 27)

2019

AMENDOLA Mario, GAFFARD Jean-Luc
Disorder and Public Concern Around Globalization

BARBAROUX Pierre
Disruptive Technology and Defence Innovation Ecosystems
(Innovation in Engineering and Technology Set – Volume 5)

DOU Henri, JUILLET Alain, CLERC Philippe
Strategic Intelligence for the Future 1: A New Strategic and Operational Approach
Strategic Intelligence for the Future 2: A New Information Function Approach

FRIKHA Azza
Measurement in Marketing: Operationalization of Latent Constructs

FRIMOUSSE Soufyane
Innovation and Agility in the Digital Age
(Human Resources Management Set – Volume 2)

GAY Claudine, SZOSTAK Bérangère L.
Innovation and Creativity in SMEs: Challenges, Evolutions and Prospects
(Smart Innovation Set – Volume 21)

GORIA Stéphane, HUMBERT Pierre, ROUSSEL Benoît
Information, Knowledge and Agile Creativity
(Smart Innovation Set – Volume 22)

HELLER David
Investment Decision-making Using Optional Models
(Economic Growth Set – Volume 2)

HELLER David, DE CHADIRAC Sylvain, HALAOUI Lana, JOUVET Camille
The Emergence of Start-ups
(Economic Growth Set – Volume 1)

HÉRAUD Jean-Alain, KERR Fiona, BURGER-HELMCHEN Thierry
Creative Management of Complex Systems
(Smart Innovation Set – Volume 19)

LATOUCHE Pascal
Open Innovation: Corporate Incubator
(Innovation and Technology Set – Volume 7)

LEHMANN Paul-Jacques
The Future of the Euro Currency

LEIGNEL Jean-Louis, MÉNAGER Emmanuel, YABLONSKY Serge
Sustainable Enterprise Performance: A Comprehensive Evaluation Method

LIÈVRE Pascal, AUBRY Monique, GAREL Gilles
Management of Extreme Situations: From Polar Expeditions to Exploration-Oriented Organizations

MILLOT Michel
Embarrassment of Product Choices 2: Towards a Society of Well-being

N'GOALA Gilles, PEZ-PÉRARD Virginie, PRIM-ALLAZ Isabelle
Augmented Customer Strategy: CRM in the Digital Age

NIKOLOVA Blagovesta
The RRI Challenge: Responsibilization in a State of Tension with Market Regulation
(Innovation and Responsibility Set – Volume 3)

PELLEGRIN-BOUCHER Estelle, ROY Pierre
Innovation in the Cultural and Creative Industries
(Innovation and Technology Set – Volume 8)

PRIOLON Joël
Financial Markets for Commodities

QUINIOU Matthieu
Blockchain: The Advent of Disintermediation

RAVIX Joël-Thomas, DESCHAMPS Marc
Innovation and Industrial Policies
(Innovation between Risk and Reward Set – Volume 5)

ROGER Alain, VINOT Didier
Skills Management: New Applications, New Questions
(Human Resources Management Set – Volume 1)

SAULAIS Pierre, ERMINE Jean-Louis
Knowledge Management in Innovative Companies 1: Understanding and Deploying a KM Plan within a Learning Organization
(Smart Innovation Set – Volume 23)

SERVAJEAN-HILST Romaric
Co-innovation Dynamics: The Management of Client-Supplier Interactions for Open Innovation
(Smart Innovation Set – Volume 20)

SKIADAS Christos H., BOZEMAN James R.
Data Analysis and Applications 1: Clustering and Regression, Modeling-estimating, Forecasting and Data Mining
(Big Data, Artificial Intelligence and Data Analysis Set – Volume 2)
Data Analysis and Applications 2: Utilization of Results in Europe and Other Topics
(Big Data, Artificial Intelligence and Data Analysis Set – Volume 3)

UZUNIDIS Dimitri
Systemic Innovation: Entrepreneurial Strategies and Market Dynamics

VIGEZZI Michel
World Industrialization: Shared Inventions, Competitive Innovations and Social Dynamics
(Smart Innovation Set – Volume 24)

2018

BURKHARDT Kirsten
Private Equity Firms: Their Role in the Formation of Strategic Alliances

CALLENS Stéphane
Creative Globalization
(Smart Innovation Set – Volume 16)

CASADELLA Vanessa
Innovation Systems in Emerging Economies: MINT – Mexico, Indonesia, Nigeria, Turkey
(Smart Innovation Set – Volume 18)

CHOUTEAU Marianne, FOREST Joëlle, NGUYEN Céline
Science, Technology and Innovation Culture
(Innovation in Engineering and Technology Set – Volume 3)

CORLOSQUET-HABART Marine, JANSSEN Jacques
Big Data for Insurance Companies
(Big Data, Artificial Intelligence and Data Analysis Set – Volume 1)

CROS Françoise
Innovation and Society
(Smart Innovation Set – Volume 15)

DEBREF Romain
Environmental Innovation and Ecodesign: Certainties and Controversies
(Smart Innovation Set – Volume 17)

DOMINGUEZ Noémie
SME Internationalization Strategies: Innovation to Conquer New Markets

ERMINE Jean-Louis
Knowledge Management: The Creative Loop
(Innovation and Technology Set – Volume 5)

GILBERT Patrick, BOBADILLA Natalia, GASTALDI Lise,
LE BOULAIRE Martine, LELEBINA Olga
Innovation, Research and Development Management

IBRAHIMI Mohammed
Mergers & Acquisitions: Theory, Strategy, Finance

LEMAÎTRE Denis
Training Engineers for Innovation

LÉVY Aldo, BEN BOUHENI Faten, AMMI Chantal
Financial Management: USGAAP and IFRS Standards
(Innovation and Technology Set – Volume 6)

MILLOT Michel
Embarrassment of Product Choices 1: How to Consume Differently

PANSERA Mario, OWEN Richard
Innovation and Development: The Politics at the Bottom of the Pyramid
(Innovation and Responsibility Set – Volume 2)

RICHEZ Yves
Corporate Talent Detection and Development

SACHETTI Philippe, ZUPPINGER Thibaud
New Technologies and Branding
(Innovation and Technology Set – Volume 4)

SAMIER Henri
Intuition, Creativity, Innovation

TEMPLE Ludovic, COMPAORÉ SAWADOGO Eveline M.F.W.
Innovation Processes in Agro-Ecological Transitions in Developing Countries
(Innovation in Engineering and Technology Set – Volume 2)

UZUNIDIS Dimitri
Collective Innovation Processes: Principles and Practices
(Innovation in Engineering and Technology Set – Volume 4)

VAN HOOREBEKE Delphine
The Management of Living Beings or Emo-management

2017

AÏT-EL-HADJ Smaïl
The Ongoing Technological System
(Smart Innovation Set – Volume 11)

BAUDRY Marc, DUMONT Béatrice
Patents: Prompting or Restricting Innovation?
(Smart Innovation Set – Volume 12)

BÉRARD Céline, TEYSSIER Christine
Risk Management: Lever for SME Development and Stakeholder Value Creation

CHALENÇON Ludivine
Location Strategies and Value Creation of International Mergers and Acquisitions

CHAUVEL Danièle, BORZILLO Stefano
The Innovative Company: An Ill-defined Object
(Innovation between Risk and Reward Set – Volume 1)

CORSI Patrick
Going Past Limits To Growth

D'ANDRIA Aude, GABARRET Inés
Building 21st Century Entrepreneurship
(Innovation and Technology Set – Volume 2)

DAIDJ Nabyla
Cooperation, Coopetition and Innovation
(Innovation and Technology Set – Volume 3)

FERNEZ-WALCH Sandrine
The Multiple Facets of Innovation Project Management
(Innovation between Risk and Reward Set – Volume 4)

FOREST Joëlle
Creative Rationality and Innovation
(Smart Innovation Set – Volume 14)

GUILHON Bernard
Innovation and Production Ecosystems
(Innovation between Risk and Reward Set – Volume 2)

HAMMOUDI Abdelhakim, DAIDJ Nabyla
Game Theory Approach to Managerial Strategies and Value Creation
(Diverse and Global Perspectives on Value Creation Set – Volume 3)

LALLEMENT Rémi
Intellectual Property and Innovation Protection: New Practices and New Policy Issues
(Innovation between Risk and Reward Set – Volume 3)

LAPERCHE Blandine
Enterprise Knowledge Capital
(Smart Innovation Set – Volume 13)

LEBERT Didier, EL YOUNSI Hafida
International Specialization Dynamics
(Smart Innovation Set – Volume 9)

MAESSCHALCK Marc
Reflexive Governance for Research and Innovative Knowledge
(Responsible Research and Innovation Set – Volume 6)

MASSOTTE Pierre
Ethics in Social Networking and Business 1: Theory, Practice and Current Recommendations
Ethics in Social Networking and Business 2: The Future and Changing Paradigms

MASSOTTE Pierre, CORSI Patrick
Smart Decisions in Complex Systems

MEDINA Mercedes, HERRERO Mónica, URGELLÉS Alicia
Current and Emerging Issues in the Audiovisual Industry
(Diverse and Global Perspectives on Value Creation Set – Volume 1)

MICHAUD Thomas
Innovation, Between Science and Science Fiction
(Smart Innovation Set – Volume 10)

PELLÉ Sophie
Business, Innovation and Responsibility
(Responsible Research and Innovation Set – Volume 7)

SAVIGNAC Emmanuelle
The Gamification of Work: The Use of Games in the Workplace

SUGAHARA Satoshi, DAIDJ Nabyla, USHIO Sumitaka
Value Creation in Management Accounting and Strategic Management: An Integrated Approach
(Diverse and Global Perspectives on Value Creation Set –Volume 2)

UZUNIDIS Dimitri, SAULAIS Pierre
Innovation Engines: Entrepreneurs and Enterprises in a Turbulent World
(Innovation in Engineering and Technology Set – Volume 1)

2016

BARBAROUX Pierre, ATTOUR Amel, SCHENK Eric
Knowledge Management and Innovation
(Smart Innovation Set – Volume 6)

BEN BOUHENI Faten, AMMI Chantal, LEVY Aldo
Banking Governance, Performance And Risk-Taking: Conventional Banks Vs Islamic Banks

BOUTILLIER Sophie, CARRÉ Denis, LEVRATTO Nadine
Entrepreneurial Ecosystems (Smart Innovation Set – Volume 2)

BOUTILLIER Sophie, UZUNIDIS Dimitri
The Entrepreneur (Smart Innovation Set – Volume 8)

BOUVARD Patricia, SUZANNE Hervé
Collective Intelligence Development in Business

GALLAUD Delphine, LAPERCHE Blandine
Circular Economy, Industrial Ecology and Short Supply Chains
(Smart Innovation Set – Volume 4)

GUERRIER Claudine
Security and Privacy in the Digital Era
(Innovation and Technology Set – Volume 1)

MEGHOUAR Hicham
Corporate Takeover Targets

MONINO Jean-Louis, SEDKAOUI Soraya
Big Data, Open Data and Data Development
(Smart Innovation Set – Volume 3)

MOREL Laure, LE ROUX Serge
Fab Labs: Innovative User
(Smart Innovation Set – Volume 5)

PICARD Fabienne, TANGUY Corinne
Innovations and Techno-ecological Transition
(Smart Innovation Set – Volume 7)

2015

CASADELLA Vanessa, LIU Zeting, DIMITRI Uzunidis
Innovation Capabilities and Economic Development in Open Economies
(Smart Innovation Set – Volume 1)

CORSI Patrick, MORIN Dominique
Sequencing Apple's DNA

CORSI Patrick, NEAU Erwan
Innovation Capability Maturity Model

FAIVRE-TAVIGNOT Bénédicte
Social Business and Base of the Pyramid

GODÉ Cécile
Team Coordination in Extreme Environments

MAILLARD Pierre
Competitive Quality and Innovation

MASSOTTE Pierre, CORSI Patrick
Operationalizing Sustainability

MASSOTTE Pierre, CORSI Patrick
Sustainability Calling

2014

DUBÉ Jean, LEGROS Diègo
Spatial Econometrics Using Microdata

LESCA Humbert, LESCA Nicolas
Strategic Decisions and Weak Signals

2013

HABART-CORLOSQUET Marine, JANSSEN Jacques, MANCA Raimondo
VaR Methodology for Non-Gaussian Finance

2012

DAL PONT Jean-Pierre
Process Engineering and Industrial Management

MAILLARD Pierre
Competitive Quality Strategies

POMEROL Jean-Charles
Decision-Making and Action

SZYLAR Christian
UCITS Handbook

2011

LESCA Nicolas
Environmental Scanning and Sustainable Development

LESCA Nicolas, LESCA Humbert
Weak Signals for Strategic Intelligence: Anticipation Tool for Managers

MERCIER-LAURENT Eunika
Innovation Ecosystems

2010

SZYLAR Christian
Risk Management under UCITS III/IV

2009

COHEN Corine
Business Intelligence

ZANINETTI Jean-Marc
Sustainable Development in the USA

2008

CORSI Patrick, DULIEU Mike
The Marketing of Technology Intensive Products and Services

DZEVER Sam, JAUSSAUD Jacques, ANDREOSSO Bernadette
Evolving Corporate Structures and Cultures in Asia: Impact of Globalization

2007

AMMI Chantal
Global Consumer Behavior

2006

BOUGHZALA Imed, ERMINE Jean-Louis
Trends in Enterprise Knowledge Management

CORSI Patrick *et al.*
Innovation Engineering: the Power of Intangible Networks

www.ingramcontent.com/pod-product-compliance
Lightning Source LLC
LaVergne TN
LVHW011916200926
848484LV00002B/640
* 9 7 8 1 7 8 6 3 0 6 4 2 5 *